Predicting Hope: Utilizing Data to End Homelessness in LA

Isaac Gyan

TABLE OF CONTENTS

ACKNOWLEDGMENTS

This book would not have been possible without the support and encouragement of many people who have helped me on this journey.

To my parents, Emmanuel and Ernestina Gyan, for instilling in me the values of hard work, resilience, and compassion. Your unwavering belief in me has been a constant source of strength.

A special thanks to Dr. Julius Agbor and Mrs. Belinda Agbor, who welcomed me with open arms and provided invaluable guidance as I adjusted to life in America. Your kindness and wisdom have had a lasting impact on my life and work.

To my mentors, colleagues, and friends who have inspired me to pursue solutions to homelessness, thank you for sharing your insights, experiences, and ideas. Your commitment to making a difference has been a driving force behind this book.

I am also grateful to the organizations and community partners who work tirelessly to support those experiencing homelessness. Your dedication to creating a better future for all motivates me to contribute to this cause.

Lastly, to every individual facing the challenges of homelessness —your stories of resilience have inspired me profoundly. This book is dedicated to you, in the hope that it can bring attention to the urgent need for lasting, data-driven solutions.

Thank you all for being a part of this journey.

Isaac Gyan

Los Angeles,California

November,2024

PREFACE

Shelter is more than just a physical structure; it is a fundamental human need—essential for survival and central to our dignity. A safe, stable home provides not only protection from the elements but also a sense of security, belonging, and personal empowerment. It is the foundation upon which we build our lives, nurture our families, and pursue our dreams.

Living in Los Angeles, I witnessed firsthand the profound impact that unstable housing can have on individuals and families. I saw how the lack of shelter could erode self-confidence, deepen poverty, and disrupt entire communities. One memory stands out: I remember a young girl, no more than 8 or 9 years old, living in a tent with her mother in a small encampment by the freeway. She had a backpack with a few worn-out books but would often look at the cars speeding by, a faraway look in her eyes as if imagining a life beyond the concrete jungle of homelessness. I often wondered: "How many others like her are there?"

Sadly, her story is not unique. According to the Los Angeles Homeless Services Authority (LAHSA), on any given night, over 66,000 people experience homelessness in LA County. Of these, nearly 17,000 are children. Homelessness in LA is not just an issue of numbers; it is a crisis that impacts families, youth, veterans, and the elderly—people from all walks of life.

Yet, amidst these staggering statistics, I also saw the resilience of the human spirit—the ability to adapt, to find hope, and to rebuild even in the most challenging circumstances. I remember speaking with a homeless man named Robert, who had been living on the streets for over five years. Despite his circumstances, he shared stories of his passion for painting and how, in moments of despair, art was his refuge. He told me, "I don't let this place break me. Every day, I remind myself that I can still create, I can still be something." His words left a deep impression on me. Robert's story, like so many others, reinforced the idea that homelessness does not define a person; it's simply a situation they are facing, one that can and should be solved.

My dedication to addressing homelessness in Los Angeles is deeply personal. LA, a city known for its diversity and innovative spirit, should be a place where everyone, regardless of their background or circumstances, has access to safe, affordable, and dignified housing. Yet, despite the city's vast resources, tens of thousands of individuals and families continue to struggle with homelessness, a crisis that seems to grow year after year. In fact, homelessness in LA has risen by 12.7% from 2020 to 2022 alone.

As a data analyst, I have spent years studying the multifaceted causes of homelessness: poverty, housing instability, mental illness, substance abuse, and systemic inequalities. Yet, even as I've worked with numbers and statistics, it is the faces behind these figures that have motivated me the most. Each statistic represents a person—someone who deserves the basic right to shelter, dignity, and respect.

In my work, I have come to understand that homelessness is not just a result of a lack of affordable housing but of a complex web of factors. Data shows that 50% of those experiencing homelessness in LA struggle with mental illness or substance abuse. But beyond the numbers, I see the human cost—the parents, children, and individuals who have been failed by a system that was meant to protect them.

The power of data in understanding and addressing homelessness cannot be overstated. Through my work, I've come to realize that data is not only an analytical tool but a transformative force for change. By utilizing data-driven approaches, we can not only identify patterns and trends but also predict where interventions will be most effective, thereby preventing homelessness before it even begins. I believe that this is the key to creating sustainable solutions that address the root causes of homelessness. A study by the National Alliance to End Homelessness found that communities that implemented data-driven initiatives saw a 20% reduction in homelessness in just two years.

My commitment to ending homelessness in LA is driven by three core values:

1. Empathy: Understanding the real, lived experience of homelessness and the desperation it brings. I think of Sarah, a mother of two, who lost her job during the pandemic and, despite her best efforts, found herself on the streets. Her story is a stark reminder of how close we all are to homelessness.

2. Justice: Recognizing the systemic failures and inequalities that contribute to and perpetuate homelessness. Black and Latino communities are disproportionately affected, making up over 60% of the homeless population in LA, despite representing only about 30% of the county's total population. These disparities highlight the deep, structural issues that need to be addressed.

3. Hope: Believing that through collective action, innovative solutions, and data-driven strategies, we can create lasting change. I have seen firsthand the transformative power of data. When we understand the trends, we can intervene early, offering resources and support before homelessness takes root.

Homelessness is not a new issue; it has been a pervasive problem for decades, with its roots embedded in a complex web of social, economic, and systemic factors. In cities across the United States, and especially in Los Angeles, homelessness continues to grow at alarming rates.

Despite the efforts of countless organizations, policymakers, and activists, the crisis remains unresolved, leaving thousands of people living on the streets, in shelters, or in temporary housing situations. But what if the solution to this crisis lies not in just providing temporary shelter, but in proactively predicting and preventing homelessness through data?

In a world where information is generated at an unprecedented rate, we often forget that behind every number, every statistic, there's a human story waiting to be told. In the case of homelessness, the numbers are staggering, but so often, the human faces behind them are obscured. Thousands of people living on the streets, in shelters, or in makeshift housing—each with a unique set of circumstances that led them to where they are today. The challenge before us is not just understanding the scale of the problem, but transforming the data into actionable insights that can lead to real, sustainable solutions.

Predicting Hope: Utilizing Data to End Homelessness is a call to action. It's about taking the data we have—sometimes overwhelming and abstract—and making it a powerful tool to end homelessness, not just in Los Angeles, but in cities across the United States and, hopefully, around the world. The approach we explore here is about more than just statistics. It's about turning data into stories, strategies, and, ultimately, lasting change.

Data has the power to make the invisible visible. It can tell us who is most at risk of becoming homeless, where they are located, and what resources they need to get back on their feet. It can tell us which services are most effective, where they are lacking, and how we can allocate resources more efficiently. Most importantly, it allows us to predict future trends, giving us the foresight to act before a crisis escalates.

This book, Predicting Hope: Utilizing Data to End Homelessness in LA, is a culmination of years of experience in business analytics, research, and a deep-seated passion for social change.

It represents my commitment to finding a sustainable solution to homelessness—one that goes beyond the reactive measures that so often dominate the conversation and focuses on a more strategic, data-driven approach. As a data analyst with over 11 years of experience across various industries, I've seen firsthand the transformative power of data in driving change. Whether it's optimizing business processes, improving financial strategies, or enhancing decision-making, data is a powerful tool. So, why not use this tool to solve one of the most pressing social issues of our time?

Los Angeles, with its sprawling neighborhoods and booming economy, is a city that should be a model of opportunity. Yet, the disparity between its wealth and the staggering number of homeless individuals tells a different story. It's a city that represents both the promise of prosperity and the stark reality of inequality. In 2020 alone, over 66,000 individuals were experiencing homelessness in LA County, with thousands more at risk of losing their homes. The traditional response to this crisis has been one of immediate intervention—providing temporary shelter, emergency services, and outreach programs. While these efforts are invaluable, they often fail to address the root causes of homelessness and, in many cases, only serve as band-aids on a much larger, systemic issue.

My approach is different. I believe we can utilize data—not just to respond to homelessness—but to predict and prevent it. Data can reveal the underlying patterns that contribute to homelessness, from rising rent prices and income inequality to mental health issues and substance abuse. By leveraging predictive analytics, we can identify individuals who are at risk of homelessness before they ever end up on the streets, allowing for early intervention that can provide stable housing and support services before the crisis deepens.

Throughout this book, you will find a combination of my professional expertise, personal passion, and a wealth of real-world case studies that demonstrate the power of data-driven solutions. The book is divided into several parts, each focusing on a different aspect of homelessness, data utilization, and how the two intersect to create meaningful, long-term solutions. In the early chapters, I will walk you through the scope of the homelessness crisis in Los Angeles, detailing the demographics, causes, and the impact it has on individuals and communities. You will gain insight into the complexity of homelessness and why traditional approaches have struggled to keep pace with the growing issue.

I'll show you how data can predict who is at the highest risk of becoming homeless, how it can identify the most effective interventions, and how we can use it to optimize the allocation of resources. We'll explore how predictive modeling, machine learning, and real-time analytics can help us make better decisions and, most importantly, help us intervene before someone falls into homelessness.

But data isn't just about numbers; it's about people. Every statistic you read in this book represents someone's story. Take, for example, the story of a woman named Maria (a composite of many real individuals I've encountered in my work). Maria had a steady job as a cashier at a local supermarket, but when her hours were reduced due to a company restructuring, she fell behind on rent. A single medical emergency—her son's asthma attack—put her further into debt, and soon, she found herself sleeping in her car with her two children.

Maria's story is one of thousands. And yet, it's a story that is often lost in the numbers. How do we prevent someone like Maria from becoming homeless? How do we intervene early, before she reaches a breaking point? This is where data comes in. By analyzing patterns in employment, housing instability, health care access, and social support systems, we can identify people at risk and provide the right kind of intervention before they fall through the cracks.

But while data is a powerful tool, it's not a magic bullet. It's not enough to simply collect and analyze data. We must also have the will to act on what we learn. We must be willing to make changes, to think creatively, and to collaborate with others. Homelessness is not just a problem for government officials or social workers to solve; it's a problem that requires the involvement of everyone—from local businesses and community organizations to residents and policymakers.

The heart of this book, however, is in its exploration of how data can be harnessed to tackle homelessness in a way that hasn't been done before. In Part II, we dive into the role of data in addressing homelessness, exploring the various sources of data—such as Housing Management Information Systems (HMIS), census data, and community-based data sources—that can be used to gain a deeper understanding of the issue. We'll look at the power of predictive modeling and data visualization, showing how these tools can help us forecast homelessness trends and allocate resources more effectively.

In Part III, we focus on the practical, actionable strategies that can be used to apply these insights. Through data-driven solutions, cities can optimize service delivery, target resources more effectively, and prevent homelessness before it happens. From identifying at-risk individuals to optimizing housing programs like Housing First and Rapid Rehousing, this section will present data-driven solutions that are already making a difference in cities across the world.

However, this book is not just about LA—it's about creating a model that can be replicated in other cities, both in the U.S. and globally. The lessons learned in Los Angeles are just the beginning. The strategies we explore here have the potential to reshape the future of homelessness prevention in any community that is committed to using data to drive change. My vision is to not only see Los Angeles become a model for ending homelessness but to inspire cities around the world to take action, using data as a tool for positive social impact.

As you read through this book, I hope you will come away with a deeper understanding of the homelessness crisis, the power of data analytics, and most importantly, the hope that meaningful change is possible. The future of homelessness doesn't have to be one of despair—it can be one of opportunity, transformation, and most of all, hope.

We are on the brink of a revolution in how we tackle social issues, and it starts with data. By making data not just a tool for business but a vehicle for social good, we can create a world where homelessness is no longer a daily reality for millions of people, but a challenge that can be solved.

Together, we can turn the tide on homelessness in LA. But we must act with urgency, creativity, and compassion. We must use data not just as a tool for analysis but as a tool for change. And most importantly, we must remember that every statistic represents a human being, a person with hopes, dreams, and the fundamental right to a safe place to call home.

As we move forward in this book, I invite you to think about the power of data. I hope you will see it not as a collection of dry numbers, but as a story of hope, resilience, and possibility. Homelessness is not just a crisis; it's an opportunity for us to reimagine how we care for one another, how we allocate resources, and how we can build a more just and equitable society.

Isaac Gyan

Los Angeles, California

November 2024

CHAPTER 1: INTRODUCTION: THE CRISIS OF HOMELESSNESS IN LA

If There's Sleepless in Seattle,
Then It's Homeless in LA!

Los Angeles, the city of dreams, innovation, and limitless opportunity, has become a paradox—a place where fame and fortune exist in close proximity to despair and homelessness. It's a city where the dazzling lights of Hollywood shine brightly, but the reality on the ground for tens of thousands of Angelenos is grim and often invisible. While many know LA for its glitzy lifestyle and iconic skyline, a darker, more urgent truth lurks in the streets, parks, and underpasses—the city's homelessness crisis.

It's as if the city, with its constant energy and drive, has forgotten that for too many, survival isn't about opportunity but about finding somewhere safe to sleep at night.

In some ways, it's fitting to draw a comparison between LA's homelessness crisis and Seattle's "Sleepless" scenario. In Seattle, the population has grown used to the endless cycle of sleepless nights, driven by relentless rain and the hustle of the city. In

LA, it's not just the sleepless nights but the homeless nights that define the experience for many. As the sun sets over the Pacific, the streets of Los Angeles come alive with those who have nowhere to go, nowhere to rest, and nowhere to turn. This is not a metaphor; it's the lived reality of nearly 66,000 men, women, and children who sleep without a roof over their heads on any given night in LA County.

This staggering number is only one part of the problem. As homelessness in LA continues to grow, the reasons behind it become more complex. Rising housing costs, economic inequality, systemic barriers, and social isolation all play their part in pushing people out of homes and into the streets. And yet, in a city that prides itself on innovation and progress, this crisis persists, year after year, showing little sign of abatement.

The Crisis in Numbers

Homelessness is a pervasive issue that challenges the very fabric of society, and nowhere is this more evident than in Los Angeles. With its sprawling neighborhoods, iconic landmarks, and vibrant culture, LA is also home to a staggering number of individuals and families experiencing homelessness. This chapter aims to provide a comprehensive overview of the crisis in Los Angeles, emphasizing the importance of addressing homelessness through innovative, data-driven approaches.

In LA, homelessness is no longer just an issue for a small segment of the population; it has become an urgent, widespread crisis that impacts every corner of the city. According to the Los Angeles Homeless Services Authority (LAHSA), more than 66,000 people in Los Angeles County are experiencing homelessness. Of those, a significant percentage—almost 75% —are living unsheltered, sleeping in tents, cars, or makeshift structures. To put this into perspective, if you lined up all the homeless individuals in LA, they would stretch from downtown to the coast, a stark visual representation of just how deep the issue runs.

What is even more troubling is that these numbers are not static. They continue to grow, with LA experiencing an 11% increase in homelessness over the past year. As the wealth gap widens and affordable housing remains out of reach for many, more and more people find themselves living on the margins. But behind these numbers lies something even more powerful—the human toll of homelessness.

The Scope of the Crisis

As of recent estimates, Los Angeles County has one of the highest rates of homelessness in the United States. Tens of thousands of people live without stable housing, facing daily struggles for survival. These individuals come from diverse backgrounds, including veterans, families with children, and those grappling with mental health issues or substance abuse. The sheer scale of the problem necessitates urgent action, as homelessness not only impacts those experiencing it but also affects the broader community.

Historical Context

Understanding the roots of homelessness in Los Angeles requires examining historical and systemic factors. The city has long been characterized by a combination of economic disparities, housing shortages, and policies that have marginalized vulnerable populations. Economic booms and subsequent busts have left many residents without the means to secure stable housing. Additionally, gentrification and rising rents have exacerbated the situation, pushing low-income families further into precarious living conditions.

The Human Side of Homelessness

Homelessness is not just a statistic; it is a human tragedy that affects people's lives in profound ways. Every person experiencing homelessness in LA has a story—a story that is too often overlooked or ignored. From veterans who once served their country to families with children who have fallen victim to rising rents, the faces of homelessness are as diverse as the city itself.

Take the story of Sofia, a single mother who once worked two jobs to support her children. After an unexpected medical emergency and the loss of her job, she was unable to keep up with her rent and found herself living in a shelter with her two young children. Or James, a talented musician who was on the verge of breaking into the industry when a series of personal setbacks left him with no choice but to live in his car. Stories like these highlight how easily a person's life can spiral into homelessness—and how often it is due to circumstances beyond their control.

But for every heartbreaking story, there is another story of resilience and strength. Many people experiencing homelessness refuse to be defined by their circumstances. They are fighting for a chance to rebuild, to reclaim their lives, and to return to their communities. Through local outreach, supportive housing programs, and innovative solutions, many have found a way out of homelessness, proving that there is hope even in the darkest of times.

Why It Persists: The Underlying Causes

Understanding the causes of homelessness in LA is key to addressing the crisis effectively. While the issue is often simplified to mental illness or addiction, the roots are far deeper and more intertwined with the city's economic and social systems. Some of the most pressing factors include:

1. Housing Affordability Crisis: Los Angeles has one of the highest costs of living in the nation, with rent prices soaring well beyond the reach of many working-class families. The shortage of affordable housing has created an environment where people are pushed into homelessness, often with little to no warning. The average rent for a one-bedroom apartment in LA is over $2,500 a month, and for many, this cost is simply unattainable.

2. Poverty and Income Inequality: The gap between the wealthy and the poor in LA is staggering. While the city boasts some of the richest individuals in the world, the disparity in income and opportunity has left many behind. According to recent reports, nearly one in five Angelenos live below the poverty line, making it difficult for them to afford basic needs, let alone a stable home.

3. Mental Health and Substance Abuse: A significant portion of the homeless population struggles with mental health disorders and substance abuse, both of which can make it difficult for individuals to maintain employment or secure housing. The lack of adequate mental health care and addiction treatment programs in LA only exacerbates these issues, leaving many vulnerable individuals without the support they need.

4. Systemic Inequality and Discrimination: Racial and ethnic disparities in LA also play a significant role in homelessness. Black, Latino, and Indigenous communities face systemic discrimination in housing, education, and employment, making them more likely to experience homelessness. Historical policies, such as redlining, continue to have lasting effects on these communities, pushing them further into poverty and instability.

5. Natural Disasters and Economic Shocks: The COVID-19 pandemic has underscored just how fragile housing stability can be. Economic downturns, natural disasters, and other unforeseen circumstances can quickly push families and individuals into homelessness. LA's increasing number of wildfires, for example, has displaced many people, while the pandemic's economic fallout has led to widespread job losses and financial insecurity.

The Human Cost

At its core, homelessness is not just a statistic; it represents human lives filled with stories of struggle, resilience, and hope. Each individual experiencing homelessness has a unique background and set of circumstances that led to their situation.

The loss of housing can result from various factors, including job loss, health crises, family breakdowns, and systemic barriers such as discrimination and lack of access to services. Recognizing the humanity behind the numbers is essential for fostering empathy and motivating action.

The Need for Innovative Solutions

Traditional approaches to addressing homelessness, such as emergency shelters and temporary housing, often fail to provide long-term solutions. While these services are crucial for immediate relief, they do not address the underlying causes of homelessness. Therefore, it is imperative to adopt more innovative, data-driven strategies that can predict trends, identify at-risk populations, and allocate resources more effectively.

In recent years, there has been a growing recognition of the role that data can play in shaping effective interventions. By leveraging data analytics and predictive modeling, stakeholders can gain insights into the dynamics of homelessness and design targeted programs that address specific needs. This shift towards evidence-based practices is essential for developing sustainable solutions that can ultimately lead to the eradication of homelessness.

The Call for Action

It's clear that the current approach to homelessness in LA is not enough. More than just temporary shelters or short-term solutions are needed. We must address the root causes of homelessness and create a systemic, long-term strategy to combat the crisis. This means using every tool available, from data analytics to innovative housing solutions, to create lasting change.

In the chapters that follow, we will delve deeper into the ways in which data and technology can transform our approach to homelessness. Through predictive analytics, resource allocation models, and policy interventions, we can create more effective strategies to reduce homelessness and ensure that all Angelenos have access to the shelter, security, and dignity they deserve.

Homelessness in LA is not just an urban problem—it's a human rights crisis. But with the right strategies, tools, and collaboration, we can begin to turn the tide and create a city where everyone has a place to call home.

As we embark on this journey to explore the intersection of data and homelessness in Los Angeles, it is crucial to remain focused on the human experiences at the heart of this issue. By understanding the scope and complexity of the homelessness crisis, we can better appreciate the potential of data to drive meaningful change. In the chapters that follow, we will delve deeper into the various facets of this issue, examining how data can be harnessed to predict hope and create a future where homelessness is no longer an inescapable reality for so many.

If there's Sleepless in Seattle, then it's Homeless in LA—but it doesn't have to be so!

ISAAC GYAN

CHAPTER 2: THE FACES OF HOMELESSNESS: DEMOGRAPHICS AND STATISTICS

Homelessness affects diverse populations, transcending age, gender, ethnicity, and socioeconomic backgrounds.

U nderstanding the demographics and statistics of homelessness in Los Angeles provides a foundation for effective solutions. By identifying who is impacted by homelessness, we can better tailor interventions to meet their unique needs. This chapter provides a detailed look at the demographics and key statistics related to homelessness in Los Angeles, shedding light on the various groups most affected and the challenges they face.

Demographics of Homelessness in LA

Homelessness in Los Angeles is not confined to any one group of people. It affects individuals across all walks of life. The following demographics help to paint a clearer picture of who is experiencing homelessness in LA.

1. Age:

26% of homeless individuals are between 25-34 years old, representing a significant portion of young adults struggling with housing insecurity.

22% are between 35-44 years old, a group that often experiences a mix of economic challenges such as job instability and rising costs of living.

15% are 55 or older, highlighting a growing trend of older adults who face homelessness due to retirement instability, health issues, and limited income.

2. Gender:

75% of the homeless population in LA is male, while 24% is female, and 1% identifies as transgender or non-binary. While men represent the majority, the challenges faced by women and gender non-conforming individuals must not be overlooked, as they experience unique vulnerabilities such as domestic violence and safety concerns.

3. Ethnicity:

43% of homeless individuals in LA are African American, which is disproportionate to their representation in the general population.

27% are Hispanic/Latino, reflecting the challenges faced by low-income communities and immigrants who often encounter barriers to affordable housing.

22% are White, indicating that homelessness is not an issue limited to any one racial group, though certain communities experience it more acutely.

6% are Asian, a minority group that is often underrepresented in discussions about homelessness but still faces unique struggles, including language barriers and lack of access to services.

4. Family Composition:

22% of homeless individuals are families with children, highlighting the growing number of children who are exposed to the instability of homelessness.

15% are single parents with children, a group that often struggles with finding affordable housing while raising children, making it harder for them to escape the cycle of homelessness.

Statistics on Homelessness in LA

The numbers surrounding homelessness in Los Angeles are staggering and provide insight into the depth of the crisis. These statistics give a snapshot of the scale of homelessness, the chronic nature of the issue, and the demographics most affected.

1. Point-in-Time (PIT) Count: According to the 2022 PIT count, 69,138 individuals experienced homelessness on a single night in Los Angeles. This number represents a snapshot of the homeless population at a given moment, but the actual number of individuals experiencing homelessness at any point throughout the year is much higher.

2. Chronic Homelessness: Of the total homeless population, 14,212 individuals have been homeless for over a year, representing those who face long-term housing instability and are often dealing with complex challenges such as mental illness and addiction.

3. Veteran Homelessness: Approximately 4,455 veterans in Los Angeles are homeless, which underscores the ongoing need to address veteran-specific issues such as PTSD, substance abuse, and reintegration into civilian life.

4. Youth Homelessness: 5,568 youth between the ages of 18-24 experience homelessness. Youth homelessness is particularly concerning because young people who experience homelessness often face a higher risk of long-term poverty, mental health

issues, and poor educational outcomes.

5. Homelessness among Seniors: In Los Angeles, 10,514 seniors aged 62 and older are homeless. This growing population of elderly homeless individuals highlights the challenges faced by older adults, such as fixed incomes, health problems, and limited access to affordable housing.

To effectively predict homelessness, several specific data sets can be utilized. These data sets provide valuable insights into the factors contributing to homelessness and help identify at-risk populations. Here are some key data sets that are particularly useful:

1. Census Data

- Description: Provides demographic information, including age, race, income, employment status, and housing conditions.

- Usefulness: Helps identify trends and disparities in housing stability across different demographic groups, which can inform targeted interventions.

2. Homeless Management Information Systems (HMIS)

- Description: A system used by service providers to collect and manage data related to homeless individuals and families.

- Usefulness: Tracks service usage, shelter stays, and demographic information, allowing for analysis of patterns and outcomes for those experiencing homelessness.

3. Social Services Data

- Description: Includes data from various social services agencies, such as food assistance programs, health services, and mental health services.

- Usefulness: Identifies individuals and families who may be at risk due to economic hardship, health issues, or lack of support systems.

4. Housing Market Data

- Description: Information on housing prices, rental rates, housing availability, and eviction rates.

- Usefulness: Analyzing trends in housing affordability can help predict increases in homelessness, particularly in areas experiencing gentrification or rapid economic change.

5. Economic Indicators

- Description: Data related to unemployment rates, poverty levels, and local economic conditions.

- Usefulness: Economic downturns often correlate with increases in homelessness; tracking these indicators can help predict potential spikes in homelessness.

6. Public Health Data

- Description: Data on health outcomes, substance abuse rates, and mental health statistics.

- Usefulness: Understanding the health issues prevalent in the homeless population can help in designing appropriate support services and preventive measures.

7. Law Enforcement and Criminal Justice Data

- Description: Information on arrests, citations, and interactions with the criminal justice system.

- Usefulness: Examining patterns of criminalization of homelessness can inform policy changes and highlight areas needing supportive services rather than punitive measures.

8. Geospatial Data

- Description: Geographic data that maps homelessness hotspots, service availability, and socio-economic conditions in different neighborhoods.

- Usefulness: Helps visualize the concentration of homelessness and identify areas where resources should be allocated.

9. Survey Data

   - Description: Data collected through surveys of homeless individuals, service providers, and community members.

   - Usefulness: Provides qualitative insights into the experiences, needs, and barriers faced by those at risk of homelessness.

10. Local and State Government Reports

   - Description: Reports from housing authorities, social service agencies, and local governments on homelessness initiatives and outcomes.

   - Usefulness: Offers insights into the effectiveness of existing programs and policies, guiding future strategies.

By integrating these diverse data sets, stakeholders can develop a comprehensive understanding of the factors contributing to homelessness in Los Angeles. This data-driven approach not only enhances the ability to predict homelessness trends but also informs the design of targeted interventions that can help prevent individuals and families from losing their homes. Through collaboration and effective data utilization, we can create a more equitable and supportive environment for all residents.

Causes of Homelessness

Understanding the root causes of homelessness is essential to finding lasting solutions. There is no single cause of homelessness, but rather a combination of factors that contribute to individuals being pushed into housing instability.

1. Poverty: An overwhelming 71% of homeless individuals cite the lack of affordable housing as the primary factor contributing to their homelessness. The affordability gap between wages and rent has become a significant barrier to housing for low-income individuals and families.

2. Unemployment: 55% of homeless individuals report losing their jobs as a major contributing factor. Job instability and the lack of well-paying, stable employment opportunities in Los Angeles create a cycle of poverty that can easily lead to homelessness.

3. Mental Illness: 44% of homeless individuals experience mental health issues, making it harder for them to find and maintain stable housing. Mental illness can interfere with individuals' ability to work, maintain relationships, and navigate bureaucratic systems to access housing and services.

4. Substance Abuse: 36% of homeless individuals struggle with addiction. Substance abuse is both a cause and a consequence of homelessness, as many individuals turn to drugs or alcohol to cope with the trauma and instability they experience while living on the streets.

Gaps in Services

Despite efforts to address homelessness in Los Angeles, significant gaps remain in the availability of essential services that can help individuals escape homelessness. These gaps often leave people without the necessary resources to regain stability and build a better future.

1. Affordable Housing: Los Angeles currently faces a shortage of 500,000 affordable housing units, making it extremely difficult for low-income individuals to secure stable housing. The skyrocketing costs of rent and limited affordable housing stock are major drivers of homelessness.

2. Mental Health Services: 40% of homeless individuals lack access to mental health services, which are critical for those suffering from conditions such as depression, PTSD, and schizophrenia. Without adequate care, individuals may be trapped in a cycle of homelessness, unable to address their underlying mental health issues.

3. Substance Abuse Treatment: 30% of homeless individuals do

not have access to substance abuse treatment, which is essential for those struggling with addiction. The lack of treatment options leaves many individuals with no means of addressing the root causes of their homelessness.

The faces of homelessness in Los Angeles reveal a complex, multifaceted issue. Demographics and statistics highlight the need for targeted solutions addressing poverty, affordable housing, mental health, and substance abuse. The crisis is not isolated to a single group of people; it is a wide-reaching problem that affects individuals across all demographics, from youth to seniors, men to women, and people of all races and ethnicities. Understanding these factors is crucial in shaping effective policy and program development that can reduce homelessness and provide individuals and families with the support they need to regain stability.

The next chapter will delve deeper into the systemic causes of homelessness in LA, including economic inequalities, social policies, and other factors that exacerbate the crisis. By understanding the broader context of homelessness, we can work towards solutions that address both the immediate needs and the long-term causes of this ongoing challenge.

CHAPTER 3: THE ROOT CAUSES OF HOMELESSNESS: POVERTY, HOUSING, AND MENTAL HEALTH

Homelessness is not an isolated issue, but rather the result of a complex interplay of factors that extend far beyond the absence of a roof over one's head.

While the immediate crisis may manifest as an individual or family lacking stable housing, the deeper causes are entrenched in social, economic, and systemic forces that affect access to housing, healthcare, employment, and mental wellness. This chapter explores the three most critical root causes of homelessness in Los Angeles: poverty, the lack of affordable housing, and mental health challenges. By examining these factors, we can understand the broader forces at play and work toward comprehensive solutions that tackle homelessness at its core.

1. Poverty: The Driving Force Behind Homelessness

Poverty is the most fundamental factor contributing to

homelessness, and it is at the root of nearly every aspect of the crisis. In Los Angeles, the gap between wages and housing costs continues to grow, leaving many individuals and families unable to afford basic necessities.

Income Inequality and Job Insecurity

Income inequality in Los Angeles is a major driver of homelessness. While the cost of living has increased dramatically, wages for low-income workers have not kept pace. The median rent for a one-bedroom apartment in LA is approximately $2,500 per month, which is out of reach for many workers who are earning minimum wage or living on a fixed income. According to the National Low Income Housing Coalition, a person would need to work nearly three full-time minimum wage jobs to afford a modest apartment in Los Angeles. For individuals working in industries such as retail, food service, and agriculture, which often provide part-time or seasonal employment with low wages, maintaining stable housing is virtually impossible.

Additionally, job insecurity exacerbates the problem. The volatility of the job market in LA, coupled with the rise of gig and freelance work, means that many people live paycheck to paycheck. One unexpected expense, such as a medical emergency or car repair, can tip the balance and leave individuals unable to pay rent. When people lose their job, their financial stability unravels quickly. With the scarcity of affordable housing, they may be forced to live in their cars, on the streets, or in temporary shelters.

Rising Income Disparity

The growing divide between the wealthy and the poor in LA is a significant factor contributing to homelessness. The tech boom and an influx of high-income professionals have driven up housing prices in the city. Meanwhile, the wages for blue-collar workers, immigrants, and people of color remain stagnated. The concentration of wealth in certain areas of the city further isolates low-income individuals from affordable housing options, and the lack of economic opportunities in marginalized communities only intensifies the crisis.

2. The Lack of Affordable Housing: A Crisis of Supply and Demand

The shortage of affordable housing is a direct result of the soaring demand for housing that far exceeds the available supply. In Los Angeles, the need for affordable housing has reached a critical level, with over 500,000 units needed to meet the demand. Despite this overwhelming need, the supply of affordable housing continues to shrink, creating an unrelenting cycle of displacement and homelessness.

A Housing Market Out of Reach

The root cause of the housing crisis in LA is simple: demand outweighs supply. The city's housing market has been priced out of reach for many working-class individuals and families. Gentrification has led to the displacement of lower-income communities, with new developments often catering to higher-income individuals. These new developments frequently do not include enough affordable units, and those that are built are often priced above what low-income tenants can afford.

A study by the California Housing Partnership Foundation found that in 2019, more than half of Los Angeles renters were spending over 30% of their income on rent, with many spending 50% or more of their income on housing. This high cost of housing makes it difficult for individuals to save money, maintain financial stability, or afford health care and other essential services. As a result, they are at constant risk of eviction or homelessness.

Evictions and the Strain on Affordable Housing

Evictions remain a significant issue in Los Angeles, particularly in the wake of the COVID-19 pandemic, when a moratorium on evictions temporarily eased the burden. According to a report by the LA Tenants Union, over 20,000 eviction notices were filed in the first year of the pandemic, even with the moratorium in place. When eviction protections are lifted, it's likely that many people will find themselves on the streets, unable to find another affordable place to live.

Furthermore, affordable housing development has been slow, despite the increasing demand. This is due, in part, to zoning regulations, insufficient funding for low-income housing, and resistance from local communities to new developments. Until the systemic barriers to affordable housing are addressed, the problem will only worsen.

3. Mental Health: A Hidden Epidemic Among the Homeless

Mental illness is a pervasive issue among the homeless population in Los Angeles, with 44% of homeless individuals reporting mental health challenges. Mental health issues are not only a consequence of homelessness but also a contributing factor to why individuals may find themselves on the streets. Mental health struggles are often exacerbated by homelessness, leading to a vicious cycle that is difficult to break.

Prevalence of Mental Illness in the Homeless Population

Mental health disorders such as schizophrenia, bipolar disorder, severe depression, and post-traumatic stress disorder (PTSD) are widespread among people experiencing homelessness. For many, untreated mental health conditions can prevent them from holding down a job, maintaining relationships, and accessing services. As a result, they may become isolated and unable to navigate the complexities of the housing market, social services, and healthcare systems. Without support, many of these individuals find themselves living in shelters or on the streets.

Barriers to Accessing Mental Health Services

Despite the high prevalence of mental illness in the homeless population, many individuals struggle to access mental health services. There is a critical shortage of psychiatric care, and the services that are available are often underfunded, difficult to access, and stigmatized. Many homeless individuals lack health insurance, making it harder to access the treatment they need. Others may face discrimination when seeking help, particularly if they have a history of substance abuse or criminal involvement. As a result, they are left untreated and often experience a worsening of their mental health conditions, which further isolates them from society and makes it more difficult to regain housing stability.

4. Addressing the Root Causes: A Comprehensive Solution

To effectively address homelessness in Los Angeles, it is essential to tackle the root causes head-on. Solutions must involve a multi-faceted approach that addresses the systemic issues of poverty, housing, and mental health.

1. Increasing Affordable Housing: This is perhaps the most urgent need in the city. Building more affordable housing units, eliminating zoning barriers, and incentivizing affordable housing development are critical steps. Moreover, rent control measures and tenant protections must be strengthened to prevent displacement.

2. Addressing Economic Inequality: Ensuring better wages and employment opportunities for low-income workers is another essential component of the solution. This includes increasing the minimum wage, creating job training programs, and improving access to stable, well-paying jobs.

3. Expanding Mental Health Services: Increasing funding for mental health programs and ensuring access to care for homeless individuals is crucial. This includes creating more mental health crisis teams, expanding community-based care, and addressing the stigma surrounding mental health treatment.

4. Ending the Cycle of Poverty: Providing more comprehensive

social services, such as food assistance, child care, and health care, would help individuals avoid falling into the trap of homelessness. Offering economic support and empowerment can help those on the brink of homelessness remain stable and housed.

The root causes of homelessness in Los Angeles are deeply interwoven, reflecting the broader issues of poverty, housing instability, and mental health challenges. Addressing these root causes requires a concerted, multi-layered effort from both the public and private sectors. By tackling these systemic issues, we can begin to stem the tide of homelessness and create a more just and equitable society. In the next chapter, we will explore the role of data in understanding homelessness and how it can be used to inform policies and initiatives aimed at ending homelessness in LA.

CHAPTER 4: CURRENT EFFORTS AND CHALLENGES: LA'S RESPONSE TO HOMELESSNESS

Los Angeles has faced the homelessness crisis with a variety of initiatives and programs over the years.

Yet despite significant financial investments, policy changes, and the collaboration of governmental and non-profit organizations, the scale of homelessness has continued to grow. The current landscape of LA's response reflects both strides in innovation and enduring obstacles. This chapter explores the main efforts undertaken by the city and county, highlights recent policies and programs, and identifies the challenges that complicate efforts to address homelessness effectively.

1. Major Initiatives and Programs

Measure H and Proposition HHH

In 2016, voters approved two landmark measures aimed at combatting homelessness in LA:

Proposition HHH provided the city with $1.2 billion to

build supportive housing, aiming to create approximately 10,000 units for homeless individuals. This funding primarily targets individuals with complex needs, including chronic homelessness, mental health conditions, and disabilities.

Measure H allocated an estimated $3.5 billion over ten years to provide supportive services, such as mental health care, substance abuse treatment, and job training. These services are designed to support the housing initiatives by helping individuals transition from the streets into stable living environments and providing the ongoing assistance needed to maintain housing.

The Los Angeles Homeless Services Authority (LAHSA)

LAHSA is a key agency in LA's homelessness response, coordinating programs and services across the county. It partners with local organizations to manage shelters, provide outreach services, and connect homeless individuals with resources. LAHSA also plays a role in the annual Point-In-Time (PIT) Count, which is instrumental in assessing the scope of homelessness and determining funding needs.

The "A Bridge Home" Program

Introduced by Mayor Eric Garcetti, the "A Bridge Home" initiative focused on establishing temporary bridge housing throughout the city. These temporary shelters provide safe, clean, and supportive spaces where individuals can stay while working with case managers to secure permanent housing. Although the initiative has expanded shelter capacity, finding appropriate locations for these shelters has been challenging due to community pushback and zoning restrictions.

Project Roomkey and Project Homekey

In response to the COVID-19 pandemic, Project Roomkey was established to provide immediate shelter in hotels and motels for homeless individuals, particularly seniors and those with underlying health conditions. This initiative reduced street homelessness and provided temporary safe havens. Building on this, Project Homekey allowed the state and local governments to acquire properties for conversion into permanent supportive housing. This program gained momentum as a rapid response to homelessness, though sustaining the funding and scaling the impact remain ongoing challenges.

Street Outreach Teams

Street outreach is an essential component of LA's homelessness response. Outreach teams—comprising social workers, medical professionals, and housing navigators—connect homeless individuals with services, distribute essential supplies, and help individuals move from the streets into shelters or transitional housing. These teams focus on high-need areas, building trust within communities and establishing pathways toward housing for individuals who may be reluctant to enter formal programs.

2. Challenges in Addressing Homelessness

Despite the efforts made, several critical challenges undermine the impact of LA's homelessness response.

Affordable Housing Shortage

The construction of affordable housing remains slow and costly. While Proposition HHH and Measure H have allocated funding for housing projects, the development process has been hindered by rising construction costs, zoning restrictions, and local opposition. Additionally, the housing produced so far has not kept pace with the need. As of recent counts, the number of affordable housing units available is only a fraction of what is required, and timelines for new projects have been extended due to budget overruns and delays.

Community Opposition (NIMBYism)

A significant barrier to expanding housing and shelter options has been the resistance from neighborhoods and communities, often termed "Not In My Backyard" (NIMBYism). Many residents are concerned that shelters or affordable housing developments will decrease property values, increase crime, or negatively impact their quality of life. This opposition has delayed or canceled numerous projects, forcing the city to consider more remote or less suitable locations for housing initiatives, which may be less accessible to critical services and resources.

Service Coordination and Funding Fragmentation

Homelessness involves multiple sectors, including housing, health, and employment. However, services often operate in silos, leading to a fragmented system where individuals fall through the cracks. Many people experiencing homelessness require a combination of housing, mental health care, addiction treatment, and job training. Coordinating these services across city, county, and non-profit agencies has proven difficult, leading to inefficiencies and missed opportunities for individuals in need.

Funding is another challenge: although initiatives like Measure H and Proposition HHH have provided significant resources, they are not enough to cover all of LA's needs. Federal and state funding supplements local efforts, but the complex grant application processes and changing policy priorities create gaps in funding continuity. As a result, certain programs experience shortages, especially for essential services like mental health care.

Mental Health and Addiction Services Shortfalls

A large proportion of homeless individuals struggle with mental health challenges or substance abuse issues, yet LA's mental health and addiction services are insufficient to meet the demand. Wait times for services can be lengthy, and there is a lack of specialized care facilities. Many shelters are not equipped to manage severe mental illness or addiction, making it difficult for some individuals to find the support they need within the existing systems. This has perpetuated a cycle of chronic homelessness, where individuals repeatedly return to the streets after failed attempts at recovery.

Bureaucracy and Policy Limitations

Bureaucratic hurdles often slow down the response to homelessness, from delays in approving housing projects to restrictions on how funds can be used. Policies intended to protect tenants or streamline processes are sometimes counterproductive. For example, rent control policies can help prevent evictions, but they can also dissuade developers from building affordable housing. Policy limitations also impact the speed at which vacant land or buildings can be converted into housing, as regulations and zoning requirements slow down or prevent the repurposing of available spaces.

3. Innovative Approaches and Promising Programs

Tiny Homes Villages

In recent years, LA has experimented with "Tiny Homes" as a way to provide transitional housing at a lower cost and with faster setup times than traditional shelters. Tiny Home Villages offer individual, lockable units, providing a greater sense of privacy and security than traditional shelters. Residents also have access to communal bathrooms, kitchens, and support services, helping them transition to more permanent housing.

Supportive Housing with Wraparound Services

Supportive housing combines permanent housing with wraparound services, such as case management, counseling, and employment support. This model addresses the needs of individuals with complex issues, helping them remain housed while receiving ongoing support. The success of supportive housing has been widely recognized, as it leads to greater stability and reduces the likelihood of individuals returning to homelessness. However, the demand for supportive housing exceeds the current supply.

Expansion of Preventative Programs

LA has increasingly focused on homelessness prevention by identifying and assisting at-risk individuals before they lose their housing. Programs that provide rental assistance, eviction defense, and emergency financial aid have been scaled up, targeting vulnerable populations who may be one crisis away from homelessness. Additionally, legal support programs for tenants facing eviction have proven effective, helping keep people housed and reducing the influx of new homeless individuals.

4. Moving Forward: Recommendations and Next Steps

While Los Angeles has made strides in its response to homelessness, there is still significant room for improvement. To make meaningful progress, several actions are recommended:

1. Expedite Housing Development: Streamlining the process for affordable housing construction and reducing zoning barriers will be essential. Expanding incentives for developers and considering adaptive reuse of existing buildings can help meet the urgent demand for affordable housing.

2. Strengthen Mental Health and Addiction Services: Increasing funding and support for mental health services will provide a more effective safety net for homeless individuals with severe mental health needs. Expanding access to community-based, culturally sensitive care is crucial to helping more people achieve stability.

3. Increase Community Engagement: Addressing community resistance to homeless services requires transparent communication, education, and community participation. Working with neighborhoods to address their concerns and ensuring new developments are well-integrated can help reduce opposition.

4. Enhance Data-Driven Policy: The integration of data

across agencies can improve the efficiency of homeless services, allowing for more targeted interventions and resource allocation. Real-time data and feedback systems will help adapt programs to meet the evolving needs of LA's homeless population.

5. Focus on Preventative Measures: Preventing homelessness before it begins is one of the most cost-effective strategies. Expanding programs that provide financial assistance, tenant protections, and job training for at-risk populations can prevent a significant number of individuals from becoming homeless.

Los Angeles has invested heavily in the fight against homelessness, yet many challenges remain. From affordable housing shortages and mental health care gaps to community resistance and funding fragmentation, the obstacles are significant. However, promising programs, innovative solutions, and increased awareness offer hope. By addressing these challenges head-on and committing to collaborative, comprehensive approaches, LA can work toward reducing homelessness and providing a path to stability for its most vulnerable residents. In the following chapter, we will examine the role of technology and data in understanding and addressing the homelessness crisis.

CHAPTER 5: THE HUMAN FACE OF HOMELESSNESS: STORIES OF STRUGGLE AND RESILIENCE

Understanding Homelessness: Definitions and Types

Homelessness is often viewed through a narrow lens, focusing primarily on individuals living on the streets or in temporary shelters. However, understanding homelessness requires a broader perspective that considers various definitions and types of homelessness. This chapter will explore the complexities of homelessness, breaking it down into distinct categories and examining the factors that contribute to each type.

Defining Homelessness

The U.S. Department of Housing and Urban Development (HUD) defines homelessness as the absence of a fixed, regular, and adequate nighttime residence. This definition encompasses several situations, including:

1. Individuals Living in Shelters: This includes emergency shelters and transitional housing programs designed to provide temporary relief.

2. Individuals Living in Places Not Meant for Human Habitation: This category includes people sleeping in cars, parks, abandoned buildings, or other locations that do not provide safe or stable living conditions.

3. Individuals Facing Imminent Homelessness: Those who are about to lose their primary housing due to eviction, foreclosure, or other circumstances.

4. Individuals Residing in Institutions: Some individuals may be temporarily housed in hospitals or correctional facilities and are likely to experience homelessness upon discharge.

Types of Homelessness

1. Chronic Homelessness

Chronic homelessness refers to individuals who have been homeless for an extended period (typically defined as a year or more) and often have a disabling condition, such as mental illness, substance use disorder, or physical disabilities. These individuals frequently cycle through emergency services, hospitals, and shelters without finding stable housing. Addressing chronic homelessness requires comprehensive support services that address both housing and health needs.

2. Episodic Homelessness

Episodic homelessness is characterized by repeated and short-lived experiences of homelessness. Individuals in this category may have temporary housing but frequently lose their stability due to various factors, such as job loss, health issues, or family disruptions. This type of homelessness often affects individuals who face economic instability or health challenges but may not have the same level of need for ongoing support as those experiencing chronic homelessness.

3. Transitional Homelessness

Transitional homelessness typically involves individuals or families who experience a brief episode of homelessness due to a specific crisis, such as job loss, divorce, or eviction. This type of homelessness is often temporary, and individuals can return to stable housing with adequate support and resources. Transitional homelessness is often a critical point where intervention can prevent longer-term homelessness.

4. Hidden Homelessness

Hidden homelessness refers to individuals who may not be counted in official homelessness statistics because they are temporarily staying with friends or family without a secure lease. This situation can create significant stress and instability for those involved, as they may be at risk of eviction from these informal arrangements. Hidden homelessness is prevalent among young people and families who lack permanent housing.

5. Family Homelessness

Family homelessness encompasses situations where entire families are without stable housing. Families may face unique challenges due to the need for stable schooling for children, access to healthcare, and the emotional stress of housing instability. Family homelessness often requires tailored interventions that consider the needs of both adults and children.

Factors Contributing to Homelessness

Understanding the various types of homelessness also involves examining the complex web of factors that contribute to housing instability. These factors include:

- Economic Factors: Rising housing costs, unemployment, and insufficient wages contribute significantly to homelessness.

- Social Factors: Family breakdowns, domestic violence, and lack of social support networks can lead individuals into precarious housing situations.

- Health Issues: Mental health disorders, substance abuse, and chronic health conditions can impede individuals' ability to maintain stable housing.

- Systemic Barriers: Discrimination, lack of access to affordable housing, and insufficient social services exacerbate the risk of homelessness for vulnerable populations.

Homelessness is a multifaceted issue that cannot be adequately addressed without a clear understanding of its definitions and types. By recognizing the diversity in homelessness experiences, stakeholders can develop targeted interventions that address the specific needs of different populations. In the following chapters, we will explore how data can be leveraged to better understand these complexities and inform effective solutions to prevent and reduce homelessness in Los Angeles.

Homelessness statistics provide essential insights into the scope of the crisis, but they rarely capture the full humanity and complexity of those living through it. Each homeless individual has a story—of dreams deferred, challenges endured, and hope that persists against the odds. In this chapter, we delve into the personal stories of people experiencing homelessness in Los Angeles, shedding light on the diverse paths that lead to the streets and the resilience that drives them to survive and strive for change.

1. The Veteran: James's Story

James served in the U.S. Army for over a decade, including two tours in Afghanistan. Upon his return, he faced the harsh reality of PTSD and the difficulty of readjusting to civilian life. Once a decorated officer, James was unprepared for the hurdles he encountered as he sought employment. The trauma he carried from war made it challenging to maintain stability in his jobs and relationships. When the combination of high rent and mounting medical bills overtook his savings, he found himself without a home.

Despite his circumstances, James maintains a sense of duty. "I may be on the streets, but I'll always be a soldier," he says. Now part of a veterans' outreach program, James advocates for mental health support and works tirelessly to help other veterans navigate homelessness. For James, the camaraderie he once found in the military now lives on in his efforts to give back, reminding him of his purpose and his identity.

2. The Single Mother: Maria's Story

Maria never imagined she would end up homeless. A mother of two young children, Maria worked tirelessly as a cleaner to make ends meet. She managed to keep her family housed until a sudden illness forced her to take unpaid leave. Without savings, she couldn't cover her rent, and eventually, her landlord evicted her.

Maria and her children spent weeks sleeping in her car until a friend referred her to a family shelter. Today, Maria is on the waiting list for transitional housing and relies on local food programs to feed her children. She dreams of building a better life, not just for herself but also to set an example for her kids. "I don't want them to grow up thinking that this is all there is," she says. Maria's resilience shines through her hardships as she clings to hope for her family's future.

3. The Youth: Jordan's Story

Jordan's story is tragically common. A young LGBTQ+ person, Jordan came out to his parents as transgender at age 17. Faced with rejection at home, he found himself on the streets. "It was like overnight, I went from being their child to a stranger," he recalls. Jordan's experience underscores the vulnerability of LGBTQ+ youth, many of whom face disproportionate rates of homelessness due to family rejection and limited support systems.

Living on the streets has exposed Jordan to dangers he never anticipated—violence, exploitation, and a lack of safe spaces. But through a local LGBTQ+ youth center, he has found support and acceptance. He now dreams of one day working to help others like himself. "I want to be the person I needed when I was alone," he says. Jordan's story highlights the unique challenges faced by LGBTQ+ homeless youth, while also illustrating the strength he has found in adversity.

4. The Senior: Diane's Story

Diane is a retired teacher who spent decades working with elementary students in LA. After her retirement, she lived on a fixed income, relying on her savings and Social Security. But as rental prices in her neighborhood skyrocketed, Diane's retirement fund dwindled faster than she had planned. When her landlord decided to raise the rent again, she found herself with no place to go.

At 67, Diane struggles with health issues that make life on the streets especially difficult. She has found a small community among other senior homeless individuals, with whom she shares resources and companionship. "You think it won't happen to you," she says, "but it can happen to anyone." Diane's story is a sobering reminder of how quickly life circumstances can change, particularly for seniors who are left vulnerable by economic pressures.

5. The Formerly Incarcerated: Marcus's Story

Marcus was released from prison two years ago after serving a sentence for a nonviolent offense. His incarceration made it difficult to find stable employment and housing, despite his desire to build a new life. Facing discrimination from landlords unwilling to rent to someone with a criminal record, Marcus found himself caught in a cycle of rejection.

With support from a local reentry program by HomelessnessinLA.com, Marcus is working toward finding permanent housing and employment. "I'm not who I was," he says, "but society keeps reminding me of who I used to be." Marcus's story reflects the uphill battle that many formerly incarcerated individuals face when trying to reintegrate, and the need for housing programs that support their reentry and growth.

6. The Mentally Ill: Sarah's Story

Sarah, once a promising student with dreams of becoming a lawyer, was diagnosed with schizophrenia in her early twenties. With limited family support, she struggled to manage her symptoms while attending school. Her condition deteriorated, and without a stable job or support network, Sarah eventually became homeless.

Today, Sarah lives in a makeshift tent near downtown Los Angeles, often experiencing episodes of paranoia and confusion. She receives periodic visits from outreach workers who provide food and medication, but consistent treatment remains out of reach. "The worst part is feeling like you don't belong anywhere," she says. Sarah's experience highlights the urgent need for comprehensive mental health services for individuals with severe mental illness.

7. The Artist: Roberto's Story

Roberto is a gifted muralist who found his way to LA with dreams of making a name in the art world. He enjoyed early success, with several commissions in the city. But a series of unfortunate events—health issues, an expensive divorce, and the loss of a major client—left him financially ruined. With no family to fall back on, he ended up on the streets.

For Roberto, art remains his lifeline. He carries a sketchpad wherever he goes, capturing moments from his life and the people he meets. Recently, a local gallery offered to showcase his work, providing a glimmer of hope. "Art keeps me alive," he says. Roberto's story reminds us of the unique talents and passions that homeless individuals bring to society and the potential that often goes unrecognized.

8. The Immigrant: Ana's Story

Ana left her home country with her husband and children in search of a better life. However, an unexpected visa issue and lack of resources left them stranded without legal status or stable employment. Her husband found work as a day laborer, but the wages were too low to cover rent. Eventually, they lost their apartment and began living in a shelter.

Despite her struggles, Ana remains hopeful. She attends English classes and is working with a legal aid organization to resolve her immigration status. "I came here so my children could have a future," she says. Ana's journey reflects the hardships faced by immigrant families and the challenges of navigating complex systems while trying to survive.

9. Resilience and Hope: A Common Thread

The stories in this chapter reveal a common thread of resilience and a will to survive against the odds. Homelessness is not an identity, but a temporary state—each of these individuals is defined not by their current situation but by their strength, dreams, and determination. From veterans and seniors to youth and families, each story highlights a different aspect of homelessness, while underscoring the shared human desire for dignity, stability, and a chance to rebuild.

Their experiences shed light on the complexity of homelessness in Los Angeles and the systemic factors that perpetuate it. These narratives are a reminder that behind each statistic is a person with a life as valuable as any other. Listening to their voices can help foster empathy, inform more humane policies, and inspire collective action.

Transforming Struggle into Advocacy

The human face of homelessness is a powerful reminder

of both the failures and the potential of our society. The strength of these individuals is matched by a growing call for action—residents, advocates, policymakers, and organizations increasingly recognize that solutions must be rooted in understanding, empathy, and inclusivity. Each story shared here is a testament to the resilience of the human spirit and serves as an urgent call to reimagine a Los Angeles where everyone has the opportunity to thrive.

CHAPTER 6. THE ROLE OF DATA IN ADDRESSING HOMELESSNESS

To effectively address homelessness,
data plays a crucial role.

I n recent years, data has emerged as a crucial tool in the fight against homelessness. By providing insights into the causes, trends, and outcomes of homelessness, data allows policymakers, social service providers, and communities to make more informed decisions.

From tracking demographics to analyzing the effectiveness of intervention programs, data can drive meaningful changes and more efficient allocation of resources. In this chapter, we explore the role of data in addressing homelessness, the various types of data collected, and the challenges and opportunities that come with harnessing data effectively.

1. Why Data Matters in Combating Homelessness

Data enables stakeholders to understand homelessness more deeply, going beyond anecdotal evidence to reveal larger patterns and trends. In Los Angeles, where homelessness is a complex and multifaceted issue, accurate data can provide insights into key questions, such as:

Who is experiencing homelessness?

Where are they located, and what are their specific needs?

What factors are most commonly associated with homelessness in LA?

Which interventions and services are proving effective, and for whom?

By answering these questions, data helps allocate resources effectively, streamline service delivery, and develop policies that address the root causes rather than merely treating the symptoms.

2. Types of Data Collected

Understanding the role of data in addressing homelessness begins with examining the various types of data collected by organizations and agencies. Each type offers unique insights and serves specific purposes in planning, policy-making, and program evaluation.

a. Point-in-Time (PIT) Counts

The Point-in-Time (PIT) count, conducted annually in many cities across the U.S., including Los Angeles, provides a snapshot of homelessness on a single night. Volunteers and outreach teams go into communities to count individuals and families experiencing homelessness. While not perfect, the PIT count offers valuable data on the total population and trends over time. However, it may underestimate numbers, as many homeless individuals avoid detection or are in places not easily accessible to counters.

b. Homeless Management Information System (HMIS)

The Homeless Management Information System (HMIS) is

a database used by agencies to track services provided to homeless individuals. It records client interactions with various services, such as shelter stays, healthcare access, and employment assistance. HMIS data enables providers to track individual progress, understand service utilization patterns, and analyze outcomes. Importantly, it also allows for cross-agency collaboration, as multiple organizations can access shared information to provide coordinated care.

c. Service and Program Data

Beyond direct homelessness data, information about housing programs, mental health services, substance abuse treatment, job training programs, and other social services is crucial. By tracking how and when individuals access these services, agencies can assess gaps in the system, identify high-need areas, and adjust services to better meet client needs.

d. Geospatial Data

Geospatial data provides insight into where homelessness is concentrated. By mapping locations where people are sleeping unsheltered, as well as the location of services, policymakers can identify geographic patterns and potential service deserts. For instance, geospatial data might reveal that certain neighborhoods lack emergency shelters or affordable housing options, guiding efforts to fill those gaps.

e. Qualitative Data

Numbers alone cannot tell the full story. Qualitative data, such as interviews, surveys, and case studies, helps capture the lived experiences of homeless individuals, offering context to the quantitative data. Stories and testimonials reveal the personal impact of policies, the specific needs of various groups, and the barriers to accessing services that people face.

3. Using Data to Drive Solutions

Data's true power lies in its ability to inform targeted, evidence-based solutions. Effective data use can improve outreach efforts, guide resource allocation, and measure the success of intervention programs. Below are key areas where data plays an essential role in creating impactful solutions.

a. Identifying High-Risk Populations

Data can help identify populations at greater risk of homelessness, such as veterans, single mothers, or young people aging out of foster care. By focusing on these groups, service providers can develop programs that address their specific needs and intervene before they experience homelessness.

b. Evaluating Program Effectiveness

Through continuous data collection and analysis, agencies can assess which programs are making a difference and which need improvement. For instance, if a job training program leads to stable employment for a significant number of participants, funding and resources can be directed toward expanding it. Conversely, programs with limited impact can be restructured or replaced.

c. Tracking Long-Term Outcomes

Long-term data allows policymakers to evaluate the effectiveness of their strategies over time. Tracking whether formerly homeless individuals remain housed and employed over months or years offers insight into program sustainability and areas for improvement.

d. Promoting Cross-Agency Collaboration

Data enables various stakeholders—such as housing authorities, healthcare providers, mental health services, and law enforcement—to work together efficiently. By sharing data, agencies can provide coordinated care, reduce service duplication, and improve overall outcomes for individuals.

e. Responding to Emerging Trends

Data helps stakeholders identify emerging trends in real-time, enabling them to respond quickly. For example, if rising rental costs are pushing more individuals into homelessness, policymakers can implement measures to address affordability. If a particular neighborhood sees an increase in homeless encampments, local agencies can deploy resources to that area.

4. Challenges in Data Collection and Use

While data is powerful, it is not without its challenges. Many of the most significant obstacles to data-driven solutions in homelessness stem from issues of accuracy, privacy, and resource allocation.

a. Data Accuracy

Collecting accurate data is challenging, as homeless individuals often move between locations or avoid detection. The PIT count, for instance, provides valuable but limited insights, and reliance solely on this data can lead to underestimation. Ensuring accuracy requires improved data collection methods and regular updates.

b. Privacy Concerns

Collecting detailed personal data raises privacy concerns, particularly in HMIS and other databases that store sensitive information. Protecting individuals' identities and maintaining ethical standards in data use are essential. Agencies must ensure that data collection complies with privacy regulations and prioritize the confidentiality of those they serve.

c. Resource Limitations

Collecting, maintaining, and analyzing data requires resources, including funding, skilled staff, and technology infrastructure. Many organizations that serve homeless populations operate on limited budgets, making it difficult to invest in robust data systems. Funding for data initiatives is essential to ensure high-quality data that can inform policies and programs.

d. Interagency Data Sharing

While collaboration is beneficial, sharing data across agencies can be complex. Differences in data collection methods, incompatible technology systems, and legal restrictions can hinder data sharing. Addressing these barriers requires standardized data collection practices and shared platforms that respect privacy and security standards.

5. Opportunities for Innovation

Despite these challenges, data presents numerous opportunities for innovation in the fight against homelessness. New technologies and methodologies can enhance data collection, analysis, and application.

a. Predictive Analytics

Predictive analytics can help identify individuals or families at risk of becoming homeless, enabling early intervention. For example, algorithms might analyze factors such as income, employment status, and healthcare needs to predict homelessness risk, allowing agencies to provide proactive support.

b. Real-Time Data Collection

Using mobile technology, social workers, outreach teams, and even homeless individuals themselves can contribute to real-time data collection. This immediacy allows service providers to respond quickly to changing conditions, such as sudden increases in unsheltered populations in particular areas.

c. Data Visualization

Data visualization tools make complex data more accessible, enabling policymakers and the public to understand trends, needs, and resource allocations at a glance. Visualizing data on maps or dashboards enhances transparency and supports informed decision-making.

d. Community-Driven Data

Involving communities in data collection empowers homeless individuals and advocates, allowing them to contribute to the information that shapes services and policies. Community-driven data initiatives foster trust and offer a more holistic view of homelessness, incorporating perspectives that may otherwise be overlooked.

6. A Data-Driven Path Forward

Data is a crucial resource in the journey to end homelessness, offering insights that lead to informed, effective interventions. In Los Angeles, where homelessness remains one of the city's most urgent challenges, leveraging data can transform how resources are allocated, how services are delivered, and how policies are crafted. With investments in data infrastructure, ethical considerations, and collaborative approaches

- Understanding the Scope: Accurate data helps policymakers understand the extent of homelessness, identifying demographics and regions most affected [1].

- Identifying Causes: Data analysis reveals root causes, such as poverty, lack of affordable housing, and mental illness, enabling targeted interventions.

- Tracking Progress: Data informs the effectiveness of existing programs, allowing for adjustments and optimization.

- Resource Allocation: Data-driven insights ensure resources are allocated efficiently, prioritizing areas of greatest need.

Some successful examples of data-driven approaches include:

- Homeless Management Information Systems (HMIS): A database tracking homelessness services, providing real-time insights.

- Geographic Information Systems (GIS): Mapping homelessness hotspots, identifying areas for targeted intervention.

- Predictive Analytics: Forecasting homelessness trends, enabling proactive responses.

By leveraging data, we can develop evidence-based solutions, ultimately reducing homelessness and providing supportive services to those in need.

CHAPTER 7: DATA SOURCES AND COLLECTION METHODS

The Data Landscape

I n the fight against homelessness, data serves as a powerful tool. It provides the evidence needed to understand the scope of the crisis, identify at-risk populations, and evaluate the effectiveness of interventions. This chapter explores the various data sources available for studying homelessness, the methodologies employed in data collection, and the challenges associated with using data effectively.

Key Data Sources

1. Census Data

The U.S. Census Bureau collects extensive demographic information every ten years, with annual updates through the American Community Survey (ACS). This data includes population demographics, economic status, housing characteristics, and geographic information.

- Usefulness: Census data helps identify trends in housing stability, income levels, and demographic shifts, laying the groundwork for understanding who is most at risk of homelessness.

2. Homeless Management Information Systems (HMIS)

HMIS are local databases that track services provided to

homeless individuals and families. These systems are used by various service providers, including shelters, housing programs, and outreach services.

- Usefulness: HMIS data includes information on service utilization, demographics, and outcomes, enabling stakeholders to analyze patterns in homelessness and the effectiveness of interventions.

3. Social Services Data

Data from social services agencies provide insights into the broader context of homelessness. This includes information from food assistance programs, mental health services, and substance abuse treatment facilities.

- Usefulness: By examining the intersection of homelessness with other social services, stakeholders can identify individuals at risk and tailor interventions to meet their needs.

4. Housing Market Data

Data on housing prices, rental rates, and eviction rates is crucial for understanding the economic factors contributing to homelessness. This information can be sourced from real estate databases, housing authorities, and local government reports.

- Usefulness: Analyzing housing market trends helps predict potential increases in homelessness, especially in areas experiencing rapid gentrification or economic downturns.

5. Economic Indicators

Economic data, such as unemployment rates, poverty levels, and wage statistics, are essential for contextualizing homelessness within broader economic trends. This information is often gathered from government agencies like the Bureau of Labor Statistics.

- Usefulness: Understanding economic conditions allows stakeholders to anticipate fluctuations in homelessness related to job markets and economic stability.

6. Public Health Data

Public health departments collect data on health outcomes, substance abuse rates, and mental health statistics. This information is vital for understanding the health challenges faced by homeless individuals.

- Usefulness: Health data can inform targeted health services and support systems, addressing the specific needs of homeless populations.

7. Geospatial Data

Geospatial data involves mapping homelessness across different neighborhoods, utilizing Geographic Information Systems (GIS) to visualize patterns and resource availability.

- Usefulness: This data helps identify homelessness hotspots, allowing for strategic resource allocation and targeted outreach efforts.

8. Survey Data

Surveys conducted with homeless individuals, service providers, and community members provide qualitative insights into the experiences and needs of those affected by homelessness.

- Usefulness: Survey data can reveal underlying issues, preferences for services, and barriers to accessing help, offering a more nuanced understanding of the homeless experience.

Methodologies for Data Collection

Quantitative Methods

Quantitative data collection involves gathering numerical data that can be statistically analyzed. This includes surveys, census data, and administrative data from HMIS. These methods allow for large-scale assessments of homelessness trends and patterns.

Qualitative Methods

Qualitative data collection focuses on understanding the experiences and perspectives of individuals. This can include interviews, focus groups, and open-ended survey questions. Qualitative methods provide depth and context to the numbers, revealing the human stories behind the statistics.

Challenges in Data Collection and Use

1. Data Fragmentation

Homelessness data is often scattered across various agencies, making it difficult to obtain a comprehensive view of the issue. Each organization may collect different types of data, leading to inconsistencies and gaps.

2. Data Privacy and Security

Data related to homelessness often contains sensitive information. Ensuring the privacy and security of this data is crucial, particularly when it involves vulnerable populations.

3. Inconsistent Definitions and Standards

Different agencies may use varying definitions of homelessness, leading to discrepancies in data collection and reporting. Establishing standardized definitions is essential for accurate analysis.

4. Underreporting

Many individuals experiencing homelessness, particularly those in hidden situations (e.g., couch surfing), may not be captured in official statistics. This underreporting can skew our understanding of the true scale of homelessness.

The data landscape surrounding homelessness is complex and multifaceted. By leveraging various data sources and methodologies, stakeholders can gain a clearer understanding of the factors contributing to homelessness and the effectiveness of interventions. However, addressing the challenges associated with data collection and use is critical to ensuring that data serves as a reliable foundation for action. In the following chapters, we will explore how predictive analytics and data-driven strategies can inform targeted interventions to prevent and reduce homelessness in Los Angeles.

Reliable data is crucial for understanding homelessness. This chapter explores data sources and collection methods, ensuring accuracy and effectiveness in addressing homelessness.

Data Sources

1. Homeless Management Information Systems (HMIS): A database tracking homelessness services.

2. Point-in-Time (PIT) Counts: Annual surveys of homeless populations.

3. Continuum of Care (CoC) Data: Regional data on homelessness services.

4. U.S. Census Bureau: Demographic data informing homelessness policies.

5. Administrative Data: Government records (e.g., healthcare, housing).

6. Surveys and Interviews: First-hand accounts from homeless individuals.

Data Collection Methods

1. Surveys and Questionnaires: Standardized tools for data collection.

2. Observational Studies: Researchers observe homelessness services.

3. Administrative Data Analysis: Examining existing government records.

4. Geographic Information Systems (GIS): Mapping homelessness hotspots.

5. Mobile Apps and Technology: Real-time data collection via apps.

Challenges and Limitations

1. Data Quality and Accuracy

2. Underreporting and Bias

3. Data Integration and Standardization

4. Privacy and Confidentiality Concerns

Best Practices

1. Collaboration and Data Sharing

2. Standardized Data Collection

3. Regular Data Updates

4. Data Analysis and Interpretation

Conclusion

Reliable data is essential for addressing homelessness. By understanding data sources and collection methods, we can develop effective solutions and policies.

CHAPTER 8: DATA VISUALIZATION AND STORYTELLING

Data visualization and storytelling play a vital role in transforming complex information into clear, compelling narratives.

This chapter explores the importance of visualizing homelessness data effectively and how storytelling can humanize statistics, making it easier for policymakers, social workers, and the public to understand and take action.

The Power of Data Visualization

Data visualization is more than just creating charts and graphs; it's about making data accessible and actionable. With the right visualization techniques, stakeholders can identify patterns, understand trends, and pinpoint the root causes of homelessness in ways that raw data alone can't accomplish. For example, a heat map showing homelessness hotspots in Los Angeles can immediately highlight areas in need of more resources or targeted interventions.

In LA, where the homeless population is diverse and widely dispersed, effective data visualization helps in:

Identifying Patterns: By visualizing factors like age, ethnicity, or duration of homelessness across different neighborhoods, trends become clearer. Visualization allows us to see, at a glance, the demographics of homelessness in LA's different regions and identify whether specific groups are more affected in certain areas.

Allocating Resources: Maps and charts help decision-makers quickly understand where resources, like shelters and outreach programs, are most needed. A bar graph showing the percentage of homeless youth, veterans, and seniors across LA, for example, could inform more balanced resource distribution.

Tracking Changes Over Time: Tracking homelessness trends over the years and visualizing this data can show whether policy interventions are effective, guiding adjustments where necessary.

Types of Data Visualization Tools for Homelessness Analysis

1. Heat Maps: These maps display the intensity of homelessness in different areas, with colors representing high- or low-density areas. In LA, heat maps can indicate neighborhoods that need urgent attention.

2. Bar Charts and Pie Charts: For breaking down demographics, such as the age, gender, and ethnicity of the homeless population, these charts make it easy to understand the composition of homelessness at a glance.

3. Line Graphs: To illustrate trends over time, like the increase or decrease in homelessness rates, line graphs provide a clear picture of the success or failure of ongoing initiatives.

4. Infographics: Infographics combine data with design to create impactful images that summarize information. For public awareness campaigns, infographics are particularly useful in showing the crisis in a way that's easy to digest.

5. Dashboards: Interactive dashboards can combine several data

points—demographics, trends, needs, and resource allocation —allowing policymakers and advocates to interact with data dynamically.

6. Story Maps: Integrating GIS data with narrative elements, story maps can tell the journey of homelessness through an area, illustrating not just data points but also connecting stories of individuals affected.

Storytelling with Data: Humanizing Homelessness

Storytelling in homelessness data goes beyond numbers. While the figures tell us how many people are homeless, they don't tell us who these people are, how they ended up in their situation, or what they're going through. This is where storytelling becomes essential.

Using personal stories and case studies can bridge the gap between data and empathy. Through storytelling, we can highlight individual narratives—showing a veteran struggling to find housing, a young mother working to break the cycle of homelessness, or a senior who lost their home due to rising rents. These stories:

Engage Emotionally: People are more likely to connect with individual stories than statistics. By putting faces to the data, we make homelessness a human issue rather than just a numerical one.

Raise Awareness and Drive Action: When shared on social media or in reports, these stories encourage the public and policymakers to act, often increasing support for homelessness initiatives.

Highlight Specific Needs: Stories can help illustrate the unique needs of different homeless groups, such as veterans needing mental health support, or families requiring childcare options.

Effective Storytelling Techniques in Homelessness Data

1. Profiles and Case Studies: Focus on real stories of

individuals and families experiencing homelessness. Highlight their backgrounds, struggles, and aspirations to give context to the data.

2. Photo Essays and Visual Narratives: Photos add an authentic, visual element to data, making the experience of homelessness more relatable. Combining images with brief descriptions or interviews can be a powerful tool for awareness.

3. Quotes and Testimonials: Including direct quotes from those impacted by homelessness can offer personal perspectives and convey the emotional reality of the statistics.

4. Interactive Stories: Websites and dashboards with interactive elements, such as clickable maps or timelines, can engage the audience more deeply and allow them to explore different aspects of the data.

5. Comparative Stories: Highlighting comparisons, such as differences in homelessness experiences among various demographic groups, can illustrate the complex, intersecting causes of homelessness.

The Role of Media and Public Advocacy

Visualizations and stories are powerful tools in media campaigns and public advocacy. Media can amplify these messages, reaching a wider audience and putting pressure on government bodies to act. Data visualizations, coupled with storytelling, allow advocacy groups to present clear, irrefutable evidence of the homelessness crisis and make a compelling case for change.

For example, a media campaign might use visual data to highlight areas of LA where the homeless population has increased most sharply over the past decade, followed by personal stories from people in those areas. This combination of data and narrative has the power to move hearts and minds, transforming public opinion and driving community involvement.

Balancing Data with Empathy

As data analysts, we must remember that behind every statistic is a real person. Data visualization and storytelling should not dehumanize those we aim to help but rather serve as tools to foster empathy and drive change. In presenting data, it's crucial to avoid reducing people to mere numbers. Instead, our work should shine a light on their stories and struggles, encouraging society to respond with compassion and action.

Data visualization and storytelling are essential tools in understanding and addressing homelessness. By making data accessible and relatable, we can turn numbers into narratives, showing the human faces behind the statistics. In doing so, we make the issue of homelessness real for people who may otherwise view it as an abstract problem, inspiring meaningful, lasting change.

In the following chapters, we will explore how to gather, analyze, and share data in a way that builds empathy, empowers stakeholders, and ultimately helps guide Los Angeles toward effective solutions for ending homelessness.

CHAPTER 9: PREDICTIVE ANALYTICS AND MODELING

Predictive analytics has the power to anticipate and address homelessness before it happens.

By analyzing existing data to identify trends, patterns, and potential outcomes, predictive modeling helps service providers, policymakers, and advocates proactively allocate resources, intervene early, and ultimately reduce homelessness in Los Angeles

Predictive analytics and modeling are crucial for proactive homelessness prevention. This chapter explores the application of advanced analytics and modeling techniques to forecast homelessness trends.

What is Predictive Analytics?

Predictive analytics is a branch of data analytics focused on making forecasts about future events based on historical data. In the context of homelessness, predictive models can estimate where and when homelessness is likely to increase, identify individuals at high risk, and suggest strategies to prevent it.

Predictive analytics includes several approaches:

Predictive Analytics Techniques

1. Regression Analysis: Identifying factors contributing to homelessness.

2. Time Series Analysis: Forecasting homelessness rates over time.

3. Machine Learning Algorithms: Predicting individual risk of homelessness.

4. Neural Networks: Identifying complex patterns in homelessness data.

5. Decision Trees and Random Forests_: Classifying high-risk populations.

Predictive analytics includes several approaches:

1. Machine Learning (ML): Using algorithms that improve over time, ML identifies patterns in large datasets and makes predictions based on these trends.

2. Statistical Modeling: Traditional methods like regression analysis identify relationships between variables, helping us understand risk factors for homelessness.

3. Neural Networks: Complex algorithms that mimic human neural networks to analyze non-linear patterns and identify hidden trends in homelessness data.

These tools enable cities like Los Angeles to shift from reactive approaches to homelessness toward proactive solutions.

Modeling Homelessness_

1. System Dynamics Modeling: Simulating homelessness systems.

2. Agent-Based Modeling: Modeling individual behaviors and interactions.

3. Microsimulation Modeling: Estimating policy impacts.

4. Spatial Modeling: Analyzing geographic factors.

Key Applications of Predictive Analytics in Homelessness

1. Identifying At-Risk Populations: Predictive models can highlight characteristics associated with higher homelessness risk, such as financial instability, health conditions, or lack of family support, allowing for targeted interventions.

2. Preventing Long-Term Homelessness: By identifying individuals at risk of chronic homelessness, predictive analytics can help service providers offer support before individuals experience long periods of instability.

3. Optimizing Resource Allocation: Predictive models can forecast where homelessness rates are likely to rise, guiding the placement of shelters, outreach programs, and other resources in areas with the greatest need.

4. Evaluating Program Effectiveness: By predicting outcomes based on current trends, these models can assess the likely impact of existing programs and inform adjustments for improved effectiveness.

5. Detecting Seasonal Trends: Predictive models can account for seasonal factors, such as increased homelessness in winter months, enabling Los Angeles to prepare with sufficient shelter and services during peak periods.

Benefits and Challenges

1. Improved forecasting and resource allocation

2. Enhanced policy evaluation and planning

3. Data quality and integration challenges

4. Model interpretability and transparency

Building a Predictive Model for Homelessness is what we intend to achieve.

Building a predictive model involves several steps, each crucial to producing accurate, actionable insights.

1. Data Collection and Preprocessing:

Data Sources: Collect data from sources like the LA Homeless Services Authority (LAHSA), census data, and health records. Combining multiple data sources provides a more comprehensive picture of homelessness.

Data Cleaning: Ensuring data is accurate, consistent, and free of errors is essential. Missing data, duplicates, and inconsistencies can lead to unreliable predictions.

Feature Engineering: This process involves identifying and creating variables that might predict homelessness, such as income level, employment status, and health conditions.

2. Choosing the Right Model:

Logistic Regression: Useful for predicting binary outcomes (e.g., whether someone will become homeless within a certain time frame).

Random Forests: An ML model that uses multiple decision trees to handle complex data and improve prediction accuracy.

Support Vector Machines (SVM): Effective for classifying data into different categories, such as "at-risk" or "low-risk" for homelessness.

Neural Networks: Especially helpful for complex relationships where traditional models fall short, although they require more data and computational power.

3. Model Training and Validation:

Training the Model: The dataset is divided into training and testing subsets. The model learns patterns in the training data to make accurate predictions.

Cross-Validation: Splitting the data further to evaluate model accuracy and prevent overfitting.

Performance Metrics: Assess the model's accuracy using metrics such as precision, recall, and F1 score to ensure reliable predictions.

4. Interpreting Model Outputs:

Predictive models can identify factors with the greatest impact on homelessness risk, informing specific intervention strategies. For example, if the model shows that a sudden drop in income increases homelessness risk, policymakers might consider programs to provide emergency rental assistance

Best Practices for Implementation

1. Collaborate with stakeholders and experts

2. Ensure data quality and standardization

3. Select suitable models and techniques

4. Monitor and refine models

Case Studies: Successful Predictive Models in Action

1. Los Angeles' Homelessness Prediction Model

2. New York City's Homelessness Prevention System

3. The Department of Veterans Affairs' Homelessness Risk Model

Several cities have effectively implemented predictive models to address homelessness. For example:

Chicago's Homelessness Prevention Model: Using historical data on evictions, health, and financial instability, Chicago developed a model to identify those at imminent risk of homelessness. This allowed for targeted outreach and early intervention, reducing new homelessness cases.

New York City's Risk Assessment Model: NYC's predictive model assesses homelessness risk among families accessing public assistance. By identifying high-risk families early, New York has provided preventative support, significantly reducing homelessness entries.

Challenges and Limitations of Predictive Analytics

While predictive analytics is powerful, it also faces limitations in the context of homelessness:

1. Data Privacy and Ethics: Homelessness data is sensitive, and maintaining privacy is critical. Collecting and using data for predictions must comply with privacy laws and ethical standards.

2. Data Quality: Predictive models rely on accurate, complete data. Inconsistent or outdated data can lead to incorrect predictions, potentially harming those the models aim to help.

3. Algorithmic Bias: Predictive models can unintentionally perpetuate biases if they're trained on historical data with embedded inequalities. For instance, a model trained on historical data may overemphasize certain demographics, resulting in unfair targeting or exclusion.

4. Interpreting Complex Models: Advanced models like neural networks can produce highly accurate predictions but may lack transparency in their decision-making process, making it harder for stakeholders to trust and understand the results.

The Future of Predictive Analytics in Homelessness Solutions

As data collection and technology evolve, predictive analytics will likely play an even larger role in homelessness prevention. Some promising developments include:

Real-Time Data Integration: Integrating real-time data, such as economic indicators or weather patterns, can improve predictive models, making them more responsive to changing conditions.

AI-Enhanced Predictions: Advances in AI, including deep learning, could enhance model accuracy and identify subtler patterns in the data, leading to more effective interventions.

Greater Collaboration: Collaborative data-sharing across government, nonprofits, and community organizations can enrich predictive models and create more comprehensive solutions.

Predictive analytics and modeling can transform homelessness prevention. By leveraging these techniques, policymakers and practitioners can develop proactive, data-driven strategies.

Predictive analytics and modeling hold enormous potential for addressing homelessness proactively. By identifying individuals and communities at risk, these models empower Los Angeles to respond before homelessness occurs, rather than reacting after the fact. This proactive approach can save resources, improve lives, and bring us closer to ending homelessness. However, as with all powerful tools, predictive analytics must be used ethically and transparently to ensure that it serves the interests of the people it seeks to help.

In the following chapters, we will explore other critical aspects of data-driven solutions for homelessness, including geospatial analysis, community outreach, and the role of cross-agency collaboration. Together, these tools offer a comprehensive framework for tackling homelessness in LA with the precision and empathy it demands.

CHAPTER 10: IDENTIFYING HIGH-RISK POPULATIONS: PREDICTIVE MODELING

Effective homelessness prevention depends on early intervention, which starts with identifying individuals and groups most at risk.

Predictive modeling is a powerful tool in this effort, enabling targeted outreach and resource allocation by uncovering patterns that signal an increased likelihood of homelessness. In this chapter, we'll dive into the strategies, data, and insights that can help stakeholders anticipate and prevent homelessness through data-driven identification of high-risk populations.

The Importance of Identifying High-Risk Populations

Identifying those most at risk allows for proactive measures that can prevent homelessness from occurring in the first place. By focusing resources on high-risk individuals, Los Angeles can reduce new homelessness cases, shorten the duration of homelessness for those affected, and allocate funds more efficiently. For example, data might indicate that individuals recently released from jail or those exiting foster care are at a heightened risk—thus enabling support networks to intervene before they lose stable housing.

Key Indicators of Homelessness Risk

Predictive models draw on various indicators to identify

individuals or groups at increased risk. These factors, often intertwined, provide early warning signals. Some of the most relevant include:

1. Economic Factors:

Income Instability: Sudden income drops or unemployment are leading indicators.

Eviction Records: Eviction history often correlates with future housing instability.

Access to Benefits: Lack of support from social services, such as unemployment or disability benefits, can contribute to risk.

2. Health and Mental Health:

Chronic Health Conditions: Individuals with chronic illnesses often face high medical expenses, leaving them financially vulnerable.

Mental Health Disorders: Mental health conditions can complicate employment and housing stability without adequate support.

Substance Use Disorders: Substance use often compounds financial and social challenges, increasing homelessness risk.

3. Social Factors:

Family Dynamics: Family breakdown, domestic violence, and lack of social support are significant contributors.

Youth Aging Out of Foster Care: Young adults who age out of foster care face a high risk due to limited resources, lack of housing stability, and insufficient support networks.

Criminal Justice System Involvement: Individuals recently released from prison or jail face difficulties securing housing, employment, and stable support networks.

4. Demographic Patterns:

Age and Gender: Studies show that younger individuals,

particularly women and LGBTQ+ youth, are at increased risk of homelessness due to discrimination and economic instability.

Veteran Status: Veterans are disproportionately represented in homelessness statistics due to challenges in reintegration, mental health conditions, and support gaps.

Building a Predictive Model for High-Risk Populations

Identifying high-risk populations through predictive modeling involves several steps:

1. Data Collection and Integration:

Integrate data from multiple sources, such as healthcare records, housing assistance data, and social services reports.

Use anonymized, aggregated data to protect individuals' privacy while still capturing trends.

2. Feature Selection:

Identify variables that are most predictive of homelessness. For example, eviction history, emergency room visits, and income fluctuations may each serve as indicators of increased risk.

Conduct exploratory data analysis (EDA) to see how these features correlate with homelessness in historical data.

3. Model Selection:

Choose a model type that balances interpretability with predictive power. Models commonly used in this context include:

Decision Trees: These models provide clear explanations of the decision-making process, making it easy to see which factors lead to a high-risk categorization.

Random Forests: By averaging multiple decision trees, these models produce more stable predictions.

Logistic Regression: Ideal for binary classifications (such as "high-risk" vs. "low-risk"), logistic regression provides an interpretable, statistically sound approach.

Neural Networks: For larger datasets, deep learning models can capture complex, non-linear relationships among variables, though they may lack transparency.

4. Testing and Validation:

Train the model on historical data, then test it on a separate dataset to ensure accuracy.

Use cross-validation to refine the model and prevent overfitting, ensuring it generalizes well to unseen data.

Evaluate the model's performance using precision, recall, and F1 scores to measure its effectiveness in identifying high-risk individuals accurately.

5. Deployment:

Collaborate with service providers to deploy the model within homelessness prevention programs.

Set up protocols for regular model updates to ensure predictions remain accurate over time as new data is collected.

Case Studies of Predictive Models in Action

Several cities have successfully implemented predictive models to preempt homelessness. For instance:

Santa Clara County: By identifying high-risk individuals in its health and human services data, Santa Clara County implemented early intervention programs, helping at-risk individuals before they reached a crisis point.

Allegheny County, Pennsylvania: The county used predictive analytics to identify families likely to face housing instability.

By providing targeted assistance, they significantly reduced new homelessness cases among these families.

Ethical Considerations in Predictive Modeling

While predictive modeling offers valuable insights, it's important to use these tools responsibly:

1. Data Privacy:

Use anonymized and aggregated data to protect individual privacy.

Ensure compliance with data protection laws, such as the California Consumer Privacy Act (CCPA).

2. Avoiding Bias and Discrimination:

Models trained on historical data can inadvertently replicate past biases (e.g., racial, socioeconomic).

Perform bias audits and adjust algorithms to mitigate potential discrimination.

3. Transparency and Trust:

Make the model's decision-making process as transparent as possible.

Communicate with communities and stakeholders about how predictive models are used, ensuring they understand and trust the process.

Challenges in Implementing Predictive Models for Homelessness

Predictive modeling also faces practical challenges that need to be addressed for effective implementation:

1. Data Gaps:

Homelessness is often a transient experience, making it hard to track individuals consistently.

Incomplete data from agencies or limited access to certain

records (e.g., health records) can hinder model accuracy.

2. Resource Limitations:

Even with predictive insights, limited resources may restrict service providers' ability to respond to all identified cases.

To maximize impact, organizations must prioritize and balance immediate interventions with long-term solutions.

3. Dynamic Nature of Homelessness:

Risk factors change over time, as do economic and social conditions. Models must be updated regularly to remain accurate.

The unpredictable nature of crises, such as natural disasters or pandemics, can suddenly alter homelessness rates.

The Future of Predictive Modeling in Homelessness Prevention

The potential for predictive modeling in homelessness prevention will likely grow as data sources expand, computing power improves, and modeling techniques evolve. Some promising trends include:

1. Real-Time Prediction Models:

Real-time data from social services, healthcare, and community organizations could enable instant identification of new risk cases, allowing for immediate intervention.

2. Cross-Sector Collaboration:

Collaboration between housing authorities, healthcare providers, and nonprofits can improve model accuracy by providing richer, more comprehensive data.

3. Advanced AI and Machine Learning:

Future models may leverage deep learning and artificial intelligence to capture complex interdependencies among risk factors, improving the accuracy of at-risk identification.

4. Customized Intervention Pathways:

Predictive models could eventually suggest tailored intervention strategies based on an individual's unique risk profile, leading to more personalized support.

Predictive modeling represents a powerful, proactive approach to tackling homelessness by identifying high-risk populations early and intervening before crisis strikes. When implemented responsibly, predictive modeling has the potential to revolutionize homelessness prevention, ensuring resources are directed toward those who need them most.

In the next chapter, we'll explore how predictive models can be combined with other data-driven strategies, such as geospatial analysis and community outreach, to create a multifaceted response to homelessness.

Through a collaborative and ethically sound approach, predictive modeling can bring Los Angeles closer to achieving a future where homelessness is both rare and brief, offering hope and stability to its most vulnerable residents.

CHAPTER 11: OPTIMIZING SERVICE DELIVERY: DATA-DRIVEN RESOURCE ALLOCATION

Effective service delivery is crucial for making meaningful strides toward alleviating the crisis.

One of the most significant advancements in recent years has been the ability to leverage data to optimize resource allocation. This chapter explores how data-driven approaches can improve the efficiency and effectiveness of services aimed at preventing and addressing homelessness in Los Angeles.

Effective resource allocation is at the heart of combating homelessness in any city, and Los Angeles is no exception. With a vast array of services—shelters, health programs, housing support, and more—optimizing the delivery of these resources is essential to maximizing impact. Data-driven resource allocation allows decision-makers to strategically distribute resources where they are most needed, ensuring that limited funds and services are used efficiently to prevent and alleviate homelessness. In this chapter, we will explore the key concepts, strategies, and tools involved in optimizing service delivery through data.

The Importance of Resource Allocation

Resource allocation is the process of distributing available resources—such as funding, staff, and services—to meet the needs of the homeless population effectively. With limited resources and a growing demand for services, it is essential to prioritize interventions that yield the greatest impact. Data-driven resource allocation helps ensure that services are not only responsive to immediate needs but also sustainable in the long term.

The Need for Optimized Resource Allocation

Los Angeles faces a homelessness crisis that is both large and complex, involving diverse populations with unique needs. To address the issue effectively, resources must be allocated to the areas where they will have the greatest impact. Without strategic planning, resources can be misdirected, leading to inefficiencies, gaps in services, and unequal access to support for vulnerable populations. Data-driven decision-making helps overcome these challenges by ensuring that the right services reach the right people at the right time.

Key Benefits of Data-Driven Resource Allocation

1. Targeting High-Need Areas: By analyzing data on homelessness trends, service providers can identify neighborhoods with the highest concentrations of homeless individuals. This targeted approach ensures that resources are allocated where they are needed most.

2. Improving Service Efficiency: Data can help streamline operations by identifying gaps in service delivery. For instance, if certain shelters are consistently at capacity while others remain underutilized, adjustments can be made to distribute resources more equitably.

3. Enhancing Program Effectiveness: By evaluating the outcomes of various programs through data analysis, stakeholders can determine which interventions are most effective. This allows for the reallocation of funds from less effective programs to those with proven success.

4. Facilitating Collaboration: Sharing data among different agencies and organizations fosters collaboration. When stakeholders have access to the same information, they can coordinate efforts and avoid duplication of services, maximizing the impact of collective resources.

Data Sources for Resource Allocation

To optimize service delivery, various data sources can be utilized:

1. Homeless Management Information Systems (HMIS)

HMIS provides detailed information about the demographics of homeless individuals, their service usage patterns, and their outcomes. This data is invaluable for understanding which services are most frequently used and which populations are underserved.

2. Geospatial Data

Geospatial analysis allows stakeholders to visualize homelessness trends across different neighborhoods. Mapping data can reveal areas with high rates of homelessness or frequent service requests, guiding resource allocation decisions.

3. Community Needs Assessments

Periodic assessments that gather input from service users, providers, and community members can help identify specific needs and gaps in services. This qualitative data complements quantitative data, providing a fuller picture of the community's needs.

4. Economic and Housing Market Data

Understanding the economic landscape—such as unemployment rates, housing availability, and rental prices—can inform decisions about where to allocate resources. Areas facing economic downturns may require more support services to prevent homelessness.

Key Data-Driven Approaches for Resource Allocation

Data-driven resource allocation relies on accurate, real-time data to guide decision-making. Some of the most powerful tools and strategies include:

1. Demand Forecasting:

Using historical data to predict the future demand for services is one of the first steps in optimizing resource allocation. By examining trends in homelessness, such as seasonal fluctuations or the impact of major events (e.g., natural disasters), city planners can anticipate needs and adjust service delivery accordingly.

For example, if data shows an uptick in homelessness during the winter months, local shelters can prepare for higher demand by increasing their capacity in advance.

2. Real-Time Data Collection and Monitoring:

Real-time data from sources such as homeless management information systems (HMIS), service providers, and emergency response teams enables city officials to track ongoing needs and adjust resources in real-time. This includes tracking available shelter beds, the availability of social services, and updates on individuals who are accessing resources.

For instance, a sudden increase in hospital visits due to a public health crisis may require the reallocation of resources to ensure homeless individuals receive appropriate medical care.

3. Geospatial Analysis:

Geographic Information Systems (GIS) tools play a critical role in optimizing resource allocation by mapping homelessness hotspots. By visualizing the locations where homeless individuals congregate or where service gaps exist, planners can identify areas where additional resources, such as shelters, healthcare services, or outreach teams, are most needed.

Geospatial analysis also helps determine which neighborhoods are most at risk of increasing homelessness, guiding investments in affordable housing, employment programs, and community services before the crisis deepens.

4. Needs-Based Assessment:

A needs-based approach involves identifying the specific requirements of different populations within the homeless community, such as families with children, veterans, individuals with mental health or substance abuse challenges, and those with chronic health conditions.

Data from assessments like the Vulnerability Index (VI-SPDAT) helps tailor services to individual needs, allowing for more effective allocation of specialized resources. For example, individuals with mental health needs may require more intensive, long-term services compared to those in need of short-term shelter or emergency relief.

5. Predictive Modeling for Service Demand:

Predictive analytics can help forecast the demand for various services based on data patterns, such as the likelihood of individuals entering homelessness based on their economic and social circumstances. By analyzing the historical paths of those who have experienced homelessness, predictive models can identify individuals at risk and guide the allocation of prevention services, such as housing subsidies, job training, and mental health support.

These models can also help forecast how resources might be distributed across the city, predicting which areas or populations will need more intensive intervention.

Implementing Data-Driven Resource Allocation

Step 1: Data Collection and Integration

The first step in optimizing resource allocation is gathering comprehensive data from multiple sources. This involves integrating data from HMIS, geospatial analyses, community surveys, and economic indicators into a centralized database.

Step 2: Data Analysis

Once the data is collected, analyzing it to identify trends and patterns is essential. Advanced analytics techniques, such as predictive modeling, can help forecast future needs based on historical data. This analysis should focus on:

- Identifying areas with rising homelessness rates.

- Understanding demographic shifts and service usage patterns.

- Evaluating the effectiveness of current programs.

Step 3: Strategic Resource Allocation

Based on the insights gained from data analysis, stakeholders can make informed decisions about resource allocation. This may involve:

- Redirecting funds to high-need areas or programs.

- Increasing capacity at underutilized shelters or services.

- Developing targeted outreach programs for specific populations, such as families or veterans.

Step 4: Monitoring and Evaluation

Continuous monitoring of resource allocation and service delivery is vital to ensure that adjustments are effective. Regularly evaluating programs through data analysis will help identify areas for improvement and ensure that resources are being used efficiently.

Strategies for Data-Driven Resource Allocation

The following strategies can be employed to enhance resource allocation and ensure that homelessness services are delivered in a timely, efficient, and equitable manner:

1. Cross-Sector Collaboration:

Effective resource allocation requires coordination among

various sectors, including healthcare, housing, law enforcement, and social services. By sharing data across these sectors, Los Angeles can ensure that resources are optimized, and no one falls through the cracks.

For example, if data shows a high number of homeless individuals with chronic medical conditions, resources can be redirected to provide more healthcare services in those areas. Similarly, information about shelter availability can be shared between different agencies to ensure that beds are used as efficiently as possible.

2. Prioritizing Resource Allocation:

Not all resources are of equal importance or urgency. A triage approach can help prioritize who gets access to resources first based on the severity of their needs. By categorizing individuals by vulnerability or urgency, agencies can direct resources to those who need them most, such as those facing imminent health crises or those with young children.

Data-driven risk assessment tools can help rank cases based on urgency, which helps service providers allocate their resources where they are most likely to have an immediate impact.

3. Cost-Benefit Analysis:

Data can be used to evaluate the effectiveness of different service models by comparing costs and outcomes. By conducting cost-benefit analyses, LA officials can determine which interventions are providing the best value for money, such as whether rapid rehousing programs or permanent supportive housing models are more effective in reducing homelessness.

These analyses can also highlight inefficiencies and areas where resources might be misallocated, such as underutilized shelters or excessive spending on services that don't yield long-term outcomes.

4. Feedback Loops and Continuous Improvement:

Continuous monitoring and evaluation of resource allocation strategies are essential for ensuring ongoing effectiveness. Data collection should be coupled with feedback mechanisms to track the outcomes of interventions and adjust them as needed.

For example, if a new housing program has been implemented in a certain area, real-time data can track its effectiveness by looking at changes in homelessness rates or client satisfaction. If the program isn't working as expected, the allocation of resources can be adjusted to improve service delivery.

5. Community Involvement:

Involving community members in the decision-making process ensures that resource allocation is responsive to local needs. By leveraging data from community engagement efforts, such as surveys or focus groups, city officials can better understand where resources are needed and tailor services to meet those needs.

Engaging individuals who have experienced homelessness in the process of resource allocation also provides invaluable insights into the barriers they face and the type of support that would be most helpful.

The Role of Technology in Optimizing Resource Allocation

Technology plays an increasingly vital role in the optimization of homelessness services. Some key technological tools include:

1. Homeless Management Information Systems (HMIS):

HMIS databases allow agencies to track service delivery across multiple providers, ensuring that resources are allocated efficiently. They provide real-time information about the availability of shelter beds, the number of clients being served, and the outcomes of various interventions.

2. Mobile Data Collection:

Mobile apps and devices allow outreach workers to collect and update data on the go, providing immediate feedback on service availability, street-level engagement, and emergency resource needs. This ensures that the data remains current and actionable, allowing for better resource management.

3. AI and Machine Learning:

Machine learning algorithms can help predict where resources are most needed by analyzing patterns in historical and real-time data. These tools can also improve decision-making processes by identifying hidden relationships within data sets, ultimately leading to more targeted interventions and more efficient use of resources.

4. Data Visualization Tools:

Visualization tools, such as dashboards and interactive maps, allow decision-makers to view and analyze data quickly, making it easier to make informed decisions about resource allocation. For example, real-time maps of shelter bed availability or outreach activities can help quickly direct teams to areas of high need.

Case Studies of Successful Data-Driven Resource Allocation

Case Study 1: Los Angeles County

In Los Angeles County, the implementation of a data-driven resource allocation strategy has led to significant improvements in service delivery. By utilizing HMIS data combined with geospatial analysis, the county identified neighborhoods with the highest rates of homelessness and tailored outreach efforts accordingly. This targeted approach resulted in increased engagement with homeless individuals and improved access to services.

Case Study 2: Seattle's Coordinated Entry System

Seattle's Coordinated Entry System uses a data-driven process to prioritize individuals for housing resources based on vulnerability and need. By analyzing data from various service providers, the system allocates resources to those with the highest acuity, ensuring that the most vulnerable populations receive immediate assistance.

Challenges in Data-Driven Resource Allocation

While data-driven resource allocation has proven to be effective, several challenges remain:

1. Data Gaps:

Incomplete or inaccurate data can compromise the effectiveness of resource allocation. For example, if there is insufficient data on the needs of specific populations, such as immigrants or individuals with disabilities, service providers may not allocate resources effectively.

2. Coordination Between Agencies:

Many homelessness service providers operate independently, which can make it difficult to share data and coordinate resource allocation. Standardized systems for data sharing and communication are needed to improve coordination between agencies.

3. Privacy and Ethical Concerns:

The collection and use of sensitive data must be handled with care to protect individuals' privacy. Transparency and clear guidelines on data use, as well as ensuring data security, are essential to gaining public trust and preventing misuse.

As we continue to explore innovative solutions, it is essential to recognize that data is not just a tool; it is a means to foster hope and create lasting change in the lives of those affected by homelessness in Los Angeles.

Optimizing service delivery through data-driven resource allocation is crucial to tackling homelessness in Los Angeles. By utilizing real-time data, predictive modeling, geospatial analysis, and needs-based assessments, city officials and service providers can direct resources more effectively, ensuring that help reaches those who need it most. Though challenges such as data gaps, coordination issues, and privacy concerns persist, the potential of data to streamline services and create long-term solutions for homelessness is undeniable. In the next chapter, we'll explore the critical role of collaboration between agencies and the community in ensuring that data-driven resource allocation translates into tangible, positive outcomes for those experiencing homelessness.

CHAPTER 12: HOUSING FIRST AND RAPID REHOUSING: DATA-DRIVEN APPROACHES

Innovative strategies have emerged
that prioritize stable housing
as a fundamental solution.

Among these, the Housing First and Rapid Rehousing models have gained significant traction. This chapter explores these approaches in depth, emphasizing how data-driven methodologies enhance their effectiveness in addressing homelessness in Los Angeles.

Addressing homelessness requires not only providing shelter but also developing sustainable solutions that allow individuals to regain their independence and stability. Among the most widely recognized and effective approaches to ending homelessness are Housing First and Rapid Rehousing. These strategies prioritize permanent housing solutions and are increasingly being enhanced by data-driven methodologies to improve their outcomes. In this chapter, we explore how Housing First and Rapid Rehousing models leverage data to provide personalized, effective support to people experiencing homelessness in Los Angeles.

Understanding Housing First

The Housing First model represents a paradigm shift in how homelessness is addressed. Unlike traditional models that require individuals to meet certain criteria (such as sobriety or employment) before receiving housing, Housing First places a high priority on providing permanent, stable housing with no preconditions. The model focuses on the idea that people are better able to address their other issues—such as substance abuse, mental health, or unemployment—once they have a secure place to live.

What is Housing First?

The Housing First model operates on the principle that providing stable, permanent housing should be the first step in addressing homelessness. This approach does not require individuals to meet certain preconditions, such as sobriety or completion of treatment programs, before obtaining housing. Instead, it focuses on providing immediate access to housing, followed by supportive services tailored to individual needs.

Core Principles of Housing First

1. Immediate Access to Housing: Individuals are placed in permanent housing as quickly as possible without any preconditions.

2. No Barriers to Entry: There are no sobriety or treatment requirements to qualify for housing.

3. Supportive Services: Once housed, individuals receive ongoing supportive services tailored to their needs, including mental health services, substance abuse treatment, and employment assistance.

4. Client-Centered: The approach is highly individualized, focusing on what the person needs to maintain housing and achieve long-term success.

Data-Driven Enhancements to Housing First

Data plays a crucial role in implementing Housing First effectively by ensuring that resources are allocated to those most in need and that support services are customized to individual situations.

1. Needs Assessment and Personalization:

Data-driven assessments, such as the Vulnerability Index – Service Prioritization Decision Assistance Tool (VI-SPDAT), are commonly used to evaluate the level of need for individuals experiencing homelessness. This tool helps identify those who are most vulnerable and in need of immediate housing.

By analyzing data on the person's health, history of homelessness, and existing support networks, service providers can develop a personalized support plan that meets the specific needs of each individual.

2. Real-Time Tracking and Case Management:

Case management software, integrated with Homeless Management Information Systems (HMIS), allows caseworkers to track the progress of individuals in the Housing First program. This system tracks metrics like housing stability, engagement in services, and improvements in health and employment.

Real-time tracking enables immediate intervention if an individual faces challenges maintaining housing, such as missed appointments or setbacks in treatment. Data allows case managers to adjust their approach quickly to prevent people from cycling back into homelessness.

3. Data-Driven Resource Allocation:

Housing First programs benefit from predictive analytics to forecast the demand for housing placements in different areas. This allows city officials to allocate housing resources more effectively, ensuring that units are available when they are most needed.

Data can also help identify areas where there is a lack of supportive services, guiding investments in mental health care, job training programs, or addiction recovery services in specific neighborhoods.

4. Evaluating Program Effectiveness:

Data is essential for evaluating the effectiveness of Housing First programs. By tracking outcomes such as housing retention rates, employment status, and improved mental health, stakeholders can identify successful interventions and areas for improvement.

Longitudinal studies that track individuals over time can provide insights into the long-term benefits of Housing First, helping to refine the model and expand its impact.

Understanding Rapid Rehousing

What is Rapid Rehousing?

Rapid Rehousing is a strategy designed to help individuals and families quickly exit homelessness and return to permanent housing. This model typically includes three key components: housing identification, rent and move-in assistance, and rapid access to supportive services.

Unlike Housing First, which aims to provide long-term housing with support services, Rapid Rehousing is a short-term intervention designed to quickly move individuals and families out of homelessness and into permanent housing. The focus of Rapid Rehousing is on securing housing as fast as possible, followed by intensive but time-limited support to help individuals achieve self-sufficiency.

Core Principles of Rapid Rehousing

1. Temporary Financial Assistance: Rapid Rehousing often includes short-term rental assistance to help individuals secure housing, with the understanding that they will eventually be able to cover their rent on their own.

2. Intensive Case Management: In addition to financial assistance, clients receive case management that focuses on increasing their ability to manage their housing, find stable employment, and access healthcare or other necessary services.

3. Time-Limited Support: Unlike Housing First, Rapid Rehousing is designed to be a short-term solution, typically providing support for 3–6 months. After this period, individuals are expected to be able to maintain housing independently.

Data-Driven Approaches in Housing First and Rapid Rehousing

The Role of Data

Data plays a crucial role in the successful implementation of both Housing First and Rapid Rehousing strategies. By leveraging data, stakeholders can enhance program effectiveness, improve service delivery, and ensure that resources are allocated efficiently.

Data-Driven Enhancements to Rapid Rehousing

1. Rapid Housing Placement with Data:

Data-driven predictive models can help caseworkers identify individuals who are most likely to succeed in a Rapid Rehousing program. By analyzing factors like employment status, health, and housing history, these models can predict which individuals will benefit most from short-term assistance.

Data analysis also helps prioritize families or individuals based on their level of vulnerability, ensuring that those who need immediate housing the most receive the fastest service.

2. Tracking Housing Stability:

Data systems track the success of Rapid Rehousing programs by measuring housing retention, employment status, and the ability to pay rent after receiving temporary financial assistance.

If individuals experience setbacks or face challenges in maintaining housing, case managers can quickly adjust support to prevent them from returning to homelessness.

3. Resource Matching and Optimization:

Real-time data on available rental units and financial

assistance programs helps caseworkers match individuals with appropriate housing options. By analyzing trends in vacancy rates and rental prices, cities can better allocate available resources and identify areas with affordable housing.

Data-driven resource allocation also allows cities to optimize the use of limited rental assistance funds, ensuring that they are directed toward individuals who need them the most.

Key Data Sources

1. Homeless Management Information Systems (HMIS): Provides detailed data on individuals accessing housing services, including demographics, service usage, and outcomes.

2. Program Evaluation Data: Collects information on the effectiveness of Housing First and Rapid Rehousing programs, tracking metrics such as housing retention rates and improvements in quality of life.

3. Community Needs Assessments: Gathers qualitative data from participants to understand their experiences and identify gaps in services.

Data-Driven Implementation

Step 1: Identifying Target Populations

Data analysis can help identify populations most in need of Housing First and Rapid Rehousing services. By examining HMIS data and community surveys, stakeholders can identify trends and demographics, ensuring that programs are tailored to meet specific needs.

Step 2: Monitoring Program Outcomes

Continuous monitoring of program outcomes is vital for assessing effectiveness. Key performance indicators (KPIs) such as:

- Housing Retention Rates: The percentage of individuals who remain in housing after six months, one year, etc.

- Income Growth: Changes in income levels among participants as they stabilize in housing.

- Service Utilization: Engagement with supportive services post-housing.

These metrics can inform adjustments to program design and service delivery.

Step 3: Evaluating Cost-Effectiveness

Data can also be used to evaluate the cost-effectiveness of Housing First and Rapid Rehousing programs. By analyzing the costs associated with providing services versus the savings generated from reduced reliance on emergency services, stakeholders can make informed decisions about funding and resource allocation.

Key Benefits of Housing First Enhanced by Data

Higher Housing Retention Rates: Data has shown that individuals in Housing First programs have higher rates of housing stability compared to those in more traditional models.

Better Health Outcomes: Data collection on health outcomes enables healthcare providers to assess the impact of stable housing on mental health, physical health, and substance use. Housing First often leads to improvements in these areas as individuals have more stable access to healthcare and are better able to focus on recovery.

Cost Efficiency: Housing First programs often reduce the long-term costs associated with homelessness, including emergency healthcare services, law enforcement interventions, and shelter costs. Data allows for the tracking of these costs and the calculation of potential savings, which can be used to advocate for continued funding and support.

Key Benefits of Rapid Rehousing Enhanced by Data

Quick Access to Housing: Rapid Rehousing focuses on placing individuals into housing quickly, a process that is expedited through data analytics.

Increased Housing Retention: By monitoring progress and providing targeted support, data helps improve housing retention rates in Rapid Rehousing programs.

Cost-Effectiveness: Rapid Rehousing can be more cost-effective than long-term shelter stays because it involves short-term assistance. Data allows for a more efficient allocation of financial resources.

Success Stories

Case Study 1: The Pathways to Housing Model

One of the most well-known examples of the Housing First approach is the Pathways to Housing model in New York City. Through rigorous data collection and analysis, the program demonstrated significant improvements in housing stability and quality of life for participants. By tracking outcomes, the program was able to refine its services and demonstrate the cost savings associated with reduced use of emergency services.

Case Study 2: Los Angeles' Rapid Rehousing Initiatives

In Los Angeles, Rapid Rehousing initiatives have shown promising results in reducing homelessness. Data from HMIS revealed that participants in these programs have higher rates of housing retention compared to those in traditional shelter systems. By continuously analyzing outcomes and participant feedback, service providers have been able to adapt their strategies to better meet the needs of homeless individuals and families.

Challenges and Considerations

1. Data Integration

One of the challenges in implementing data-driven approaches is the integration of data from various sources. Ensuring that information from HMIS, service providers, and community assessments is harmonized is essential for accurate analysis.

2. Privacy Concerns

As with any data collection involving vulnerable populations, privacy and security are paramount. Programs must develop robust protocols to protect participant information while still utilizing data for program improvement.

3. Ensuring Accessibility

Data-driven approaches must consider the accessibility of services. Ensuring that housing options and supportive services are available to all individuals, regardless of their circumstances, is crucial for the success of these models.

Comparing Housing First and Rapid Rehousing

Both Housing First and Rapid Rehousing are effective models for addressing homelessness, but they serve different purposes and populations. While Housing First focuses on long-term stability and comprehensive support, Rapid Rehousing aims to quickly end homelessness with time-limited support. Data enhances both models by ensuring that resources are directed where they are most needed, and outcomes can be tracked and evaluated to improve long-term solutions.

Housing First	Rapid Rehousing
Focus on Permanent Housing first	Focus on quick housing placement with short term assistance.

No preconditions for entry	Time limited assistance with a focus on self sufficiency.
Long term client centered support	Short term intensive case management.
Provides stability through permanent housing .	Provides short term financial assistance and case management.
Enhanced by long term tracking and data analysis.	Enhanced by real time data and predictive modeling for housing placement.

Housing First and Rapid Rehousing represent transformative approaches to addressing homelessness, grounded in the understanding that stable housing is a fundamental human right. By utilizing data-driven strategies, stakeholders can optimize these models, ensuring that they are effective, efficient, and responsive to the needs of the community.

Both Housing First and Rapid Rehousing are integral to solving homelessness in Los Angeles, but they are most effective when enhanced by data. By using data to track needs, prioritize placements, monitor progress, and adjust interventions, these models become even more impactful in helping individuals regain housing stability. As we continue to refine these approaches, data will play an increasingly critical role in ensuring that homelessness is not only addressed but ultimately eradicated in our communities.

As Los Angeles continues to confront the challenges of homelessness, the integration of data into these approaches will be vital in creating lasting solutions that empower individuals and families to reclaim their lives and thrive in stable housing.

CHAPTER 13: ADDRESSING MENTAL HEALTH AND SUBSTANCE ABUSE: DATA-DRIVEN SOLUTIONS

The intersection of homelessness with mental health and substance abuse is a critical issue that demands an integrated, data-driven approach.

These two factors are often significant contributors to an individual's experience of homelessness, and without addressing them effectively, long-term stability is hard to achieve. Data-driven solutions have emerged as a transformative tool in identifying, diagnosing, and managing mental health and substance use disorders within the homeless population. This chapter explores the role of data in addressing these intertwined issues and presents strategies to improve outcomes for people experiencing homelessness in Los Angeles.

Mental health and substance abuse are critical issues that significantly contribute to homelessness. Individuals facing these challenges often find it difficult to maintain stable housing, creating a cycle that can be hard to break. This chapter explores the intersection of homelessness, mental health, and substance abuse, emphasizing data-driven solutions that can help address these complex issues in Los Angeles.

Understanding the Scope of the Issues

Mental Health and Homelessness

Mental health disorders are prevalent among homeless populations, with studies indicating that a significant percentage of individuals experiencing homelessness also suffer from conditions such as depression, anxiety, bipolar disorder, and schizophrenia. These mental health challenges can hinder one's ability to secure and maintain housing, navigate social services, and engage in employment.

The Role of Mental Health and Substance Abuse in Homelessness

Mental health disorders and substance abuse are highly prevalent among people experiencing homelessness. Studies have shown that up to 25-30% of the homeless population struggles with severe mental illnesses, such as schizophrenia, bipolar disorder, and depression. In addition, substance abuse —including alcohol and drugs—is often either a cause or a consequence of homelessness, exacerbating an individual's vulnerability.

The Complex Relationship:

1. Mental Health and Homelessness:

People with severe mental health conditions may struggle to secure and maintain housing due to a lack of stable income, difficulty in managing daily living tasks, and poor physical health. Mental health issues can lead to social isolation and make it difficult for individuals to adhere to traditional housing models that require independence.

Mental health disorders can also make individuals more susceptible to becoming homeless in the first place, especially if there is a lack of family support or access to treatment services.

2. Substance Abuse and Homelessness

Substance abuse is another major factor contributing to homelessness. Many individuals struggle with addiction to alcohol or drugs, which can lead to financial instability, relationship breakdowns, and legal issues. Substance abuse often co-occurs with mental health disorders, complicating treatment and recovery efforts.

Many individuals experiencing homelessness turn to substances as a way to cope with trauma, anxiety, or depression. In some cases, substance abuse is linked to the breakdown of familial and social ties, leading to homelessness.

Substance use disorders also complicate efforts to exit homelessness, as individuals often face discrimination from landlords or service providers, which can further entrench their unstable living situations.

The Importance of Integrated Approaches

Addressing the intertwined issues of mental health and substance abuse requires integrated approaches that combine housing support with mental health and addiction services. This holistic method recognizes that stable housing can serve as a foundation for recovery, enabling individuals to access necessary treatment and support.

To break the cycle of homelessness, it's imperative to provide integrated services that address both mental health and substance abuse, alongside housing solutions.

The Role of Data in Developing Solutions

Key Data Sources

To effectively address mental health and substance abuse among homeless populations, various data sources can be utilized:

1. Homeless Management Information Systems (HMIS): Provides insights into the demographics and service utilization of homeless individuals, including mental health and substance abuse issues.

2. Public Health Data: Offers information on the prevalence of mental health disorders and substance abuse within

the community, helping to identify trends and high-risk populations.

3. Service Utilization Data: Tracks the engagement of homeless individuals with mental health and substance abuse services, allowing for analysis of access and outcomes.

4. Community Surveys: Collect qualitative data on the barriers individuals face in accessing mental health and substance abuse treatment, providing insights into their specific needs.

Data-Driven Strategies

Step 1: Identifying At-Risk Populations

Data analysis can help identify individuals at higher risk of homelessness due to mental health or substance abuse issues. By examining HMIS and public health data, stakeholders can pinpoint trends and demographics, allowing for targeted outreach and intervention strategies.

Step 2: Enhancing Service Coordination

Effective coordination among mental health, substance abuse, and housing service providers is essential. Data integration allows for the sharing of information across agencies, ensuring that individuals receive comprehensive support. Coordinated entry systems can help streamline access to services, reducing barriers to treatment.

Step 3: Monitoring Outcomes

Continuous monitoring of program outcomes is vital for assessing the effectiveness of integrated approaches. Key performance indicators (KPIs) might include:

- Housing Retention Rates: The percentage of individuals with mental health or substance abuse issues who maintain stable housing.

- Treatment Engagement: The rate at which individuals access and participate in mental health or substance abuse treatment after securing housing.

- Quality of Life Improvements: Assessing changes in individuals' overall well-being, including mental health status and substance use patterns.

Evaluating these outcomes can inform adjustments to programs and services, ensuring they are responsive to the needs of participants.

Data-Driven Approaches to Addressing Mental Health and Substance Abuse

The power of data lies in its ability to inform personalized, evidence-based interventions. By leveraging data to better understand the challenges faced by individuals experiencing homelessness, service providers and policymakers can tailor their strategies to improve access to care, enhance recovery outcomes, and prevent homelessness from recurring.

1. Early Identification and Screening

Data-driven tools are essential in identifying individuals at risk of mental health or substance use disorders before they fall into homelessness. Predictive analytics and screening tools can be used to assess risk factors, such as trauma history, family background, and socio-economic conditions, which may lead to mental health or substance abuse issues.

Screening Tools: Standardized screening tools like the PHQ-9 for depression or the AUDIT-C for alcohol use disorder can be integrated into intake forms for shelters, service providers, and outreach teams. These tools can identify individuals who may need further assessment and targeted support.

Predictive Analytics: By analyzing patterns of behavior and background information, predictive models can help case managers identify which individuals are more likely to experience mental health crises or substance abuse problems. This allows for earlier intervention and access to appropriate services.

2. Integrated Service Models

One of the most promising data-driven approaches to addressing homelessness-related mental health and substance abuse issues is the integrated service model. This model combines mental health, substance abuse, and housing services under one roof, ensuring that individuals receive holistic support to address the root causes of their homelessness.

Coordinated Care Systems: Using data, service providers can establish coordinated care systems that track individuals' progress across multiple service providers. This ensures that all aspects of an individual's recovery, including mental health treatment, substance use recovery, and housing stability, are monitored and managed.

Data Sharing Across Systems: Information from hospitals, shelters, substance abuse clinics, and mental health facilities can be shared securely between service providers, enabling a more efficient and effective treatment plan. Data-sharing platforms allow case managers to monitor a client's needs, track medication regimens, and assess the overall progress of their rehabilitation.

3. Real-Time Monitoring and Case Management

Using Homeless Management Information Systems (HMIS) and case management software, agencies can collect and analyze real-time data about an individual's status, including mental health evaluations, treatment participation, substance use patterns, and housing stability.

Data-Driven Case Management: Caseworkers can track the progress of clients and intervene quickly when necessary. For example, if a client is showing signs of relapsing or experiencing a mental health crisis, the system can alert the caseworker to take immediate action, whether that be through additional counseling, hospitalization, or adjusting the individual's housing support.

Client-Level Data: Collecting data at the individual level allows service providers to provide a personalized approach to care. This data includes mental health diagnoses, substance use history, current medications, treatment plans, and housing preferences, allowing caseworkers to understand the complexity of the individual's needs and make more informed decisions.

4. Tracking Recovery and Long-Term Outcomes

Data can also be used to track recovery outcomes over the long term. By collecting data over time on mental health stability, substance use recovery, and housing retention, agencies can measure the success of their interventions and adapt strategies accordingly.

Longitudinal Data: Tracking individuals over time can help identify trends and challenges in maintaining housing stability, particularly when mental health or substance abuse issues are involved. By collecting longitudinal data, organizations can evaluate whether interventions lead to long-term improvements and refine their strategies to maximize success.

Program Evaluation: Data allows organizations to evaluate the effectiveness of different mental health and substance abuse treatments in combination with housing solutions. By analyzing these outcomes, cities can determine which models are most effective at reducing recidivism and improving housing stability.

5. Resource Allocation and Service Optimization

Data can be used to optimize the allocation of resources, ensuring that individuals with the most urgent mental health or substance abuse needs receive immediate attention.

Predictive Modeling for Resource Allocation: Predictive analytics can help direct resources to areas with the greatest demand for mental health and substance abuse services, enabling agencies to allocate support efficiently. This data-driven approach can ensure that resources are used where they will have the most significant impact.

Mapping Service Gaps: Geospatial data can be used to map areas with unmet needs for mental health and substance abuse services. By identifying where gaps exist, cities can better direct funding to regions with the most need.

Case Studies of Data-Driven Solutions

Case Study 1: The PATH (People Assisting the Homeless) Model

PATH in Los Angeles employs a data-driven approach to integrate housing and supportive services for individuals facing homelessness due to mental health and substance abuse issues. By utilizing HMIS data, PATH identifies individuals in need of immediate housing and connects them to appropriate mental health and addiction services. The program has demonstrated significant improvements in housing retention and overall well-being among participants.

Case Study 2: Project Roomkey

In response to the COVID-19 pandemic, Los Angeles launched Project Roomkey, a data-driven initiative aimed at providing temporary housing for vulnerable populations, including those with mental health and substance abuse challenges. By analyzing data on homelessness and health risks, the city was able to prioritize individuals for housing and connect them to essential services. The project showcased the effectiveness of rapid response strategies informed by data.

Challenges and Considerations

1. Data Privacy and Security

When dealing with sensitive information related to mental health and substance abuse, ensuring data privacy and security is essential. Programs must implement robust protocols to protect participant information while still leveraging data for improvement.

2. Stigma and Engagement

Stigma surrounding mental health and substance abuse can hinder individuals from seeking help. Data-driven outreach strategies must consider these barriers and work to build trust within the community, encouraging individuals to access services.

3. Funding and Resource Allocation

Integrated approaches often require additional funding and resources. Data can help demonstrate the effectiveness of these strategies, supporting efforts to secure funding from government and philanthropic sources.

Ethical Considerations in Data Collection

While data plays an essential role in addressing mental health and substance abuse in the homeless population, it also raises important ethical considerations.

1. Privacy and Consent: People experiencing homelessness are often highly vulnerable, and their personal information must be protected. Data collection should always be done with the informed consent of the individual, ensuring they are aware of how their information will be used.

2. Data Security: Sensitive health information must be stored and shared securely to prevent unauthorized access. This is especially important when dealing with vulnerable populations, whose data could be misused or exposed.

3. Avoiding Stigmatization: The data collected should be used to help individuals, not to label or stigmatize them. Programs must be designed to ensure that individuals are treated with dignity and that the data is used to enhance care, rather than perpetuating negative stereotypes.

Data-driven solutions offer a powerful tool for addressing

the mental health and substance abuse challenges faced by individuals experiencing homelessness. By leveraging predictive analytics, real-time case management, integrated service models, and resource optimization, we can ensure that people receive the targeted support they need to overcome these complex issues and regain stability. As Los Angeles and other cities continue to innovate in this area, it is crucial to remember that data alone is not enough—it must be combined with compassionate care and a commitment to addressing the underlying social issues that drive homelessness.

Addressing mental health and substance abuse among homeless populations is critical to breaking the cycle of homelessness. By employing data-driven solutions, stakeholders can identify at-risk individuals, enhance service coordination, and monitor outcomes to ensure that interventions are effective. As Los Angeles continues to grapple with homelessness, the integration of mental health and substance abuse services into housing initiatives will be essential in creating pathways to stability and recovery for those in need. Through collaboration, innovation, and a commitment to data-driven practices, we can foster hope and create lasting change in the lives of vulnerable individuals and families.

CHAPTER 14. ENGAGING WITH HOMELESSNESS PREVENTION: EARLY INTERVENTION AND OUTREACH

Homelessness is a complex issue affecting millions worldwide.

Preventing homelessness requires proactive and comprehensive approaches, emphasizing early intervention and outreach. This essay explores the importance of engaging with homelessness prevention through early intervention and outreach.

Preventing homelessness before it occurs is a critical strategy in addressing the crisis effectively. Early intervention and outreach play pivotal roles in identifying at-risk individuals and families, providing them with the necessary support to maintain their housing stability. This chapter explores the importance of proactive measures, effective outreach strategies, and data-driven approaches to homelessness prevention in Los Angeles.

This chapter also explores how early intervention and outreach efforts can be enhanced through the use of data, and the vital role these strategies play in addressing homelessness in Los Angeles.

The Importance of Homelessness Prevention

Early intervention refers to the proactive identification and assistance provided to individuals who are at risk of becoming homeless. Research has shown that the earlier individuals receive support, the more likely they are to retain housing and avoid the cycle of homelessness. Many individuals do not suddenly become homeless; instead, they experience a series of escalating challenges—financial instability, job loss, family conflict, mental health or substance abuse issues—that increase their vulnerability.

By addressing these issues early, communities can prevent the escalation that leads to homelessness. Early intervention may involve providing housing subsidies, legal assistance, mental health counseling, or job training. It is a compassionate approach that not only improves outcomes for individuals but also saves significant public resources by reducing the need for emergency shelters, healthcare services, and long-term social services.

Understanding the Cost of Homelessness

Addressing homelessness after it occurs is significantly more costly than implementing preventive measures. Homelessness often leads to increased reliance on emergency services, healthcare, and law enforcement, placing a financial burden on communities. By investing in prevention strategies, cities can reduce these costs while enhancing the overall well-being of their residents.

Early Intervention

Early intervention is crucial in preventing homelessness. Identifying at-risk populations and providing supportive services prevents individuals from slipping into homelessness. Rapid rehousing programs, emergency financial assistance, and mental health and substance abuse services address underlying issues.

Examples:

1. Los Angeles' Homeless Prevention Program provides emergency financial assistance and rapid rehousing support.

2. New York City's Homebase program offers case management and financial assistance.

Key Aspects of Early Intervention:

1. Identification of At-Risk Populations: Utilizing data to identify individuals who are at a high risk of becoming homeless.

2. Targeted Support Services: Providing timely interventions like financial assistance, family mediation, job placement, and mental health counseling.

3. Engagement with Vulnerable Groups: Focusing on groups that are more likely to face homelessness, such as youth aging out of foster care, veterans, or individuals exiting the criminal justice system.

The Benefits of Early Intervention

Early intervention refers to identifying individuals and families at risk of homelessness before they lose their housing. This proactive approach can include:

- Financial Assistance: Providing temporary financial aid for rent or utilities to prevent eviction.

- Supportive Services: Offering case management, counseling, and connections to resources that address underlying issues such as job loss, health crises, or family disputes.

- Education and Awareness: Informing at-risk populations about available resources and support services to empower them to seek help before facing housing instability.

Outreach

Outreach is the process of actively engaging individuals who may be at risk of homelessness but have not yet reached a crisis point. Outreach workers, often employed by homeless service agencies, visit locations such as shelters, community centers, hospitals, and even street encampments, to connect individuals with resources and services that can help them stay housed.

The goal of outreach is not only to offer immediate help but to build trust and rapport with individuals, many of whom may have become disillusioned with or distrustful of service systems. By offering ongoing support and offering services before the person loses their home, outreach efforts can provide a critical lifeline.

Effective Outreach Strategies

Building Trust in the Community

Engagement with at-risk populations requires building trust and rapport within the community. Outreach teams should prioritize establishing relationships with individuals and families, particularly those who may be hesitant to seek help due to stigma or past negative experiences with service providers.

Effective Outreach Strategies:

1. Proactive Engagement: Rather than waiting for individuals to seek help, outreach workers identify those who are struggling and initiate contact.

2. Building Trust: Given the skepticism and trauma that many individuals experiencing homelessness face, building relationships based on trust is essential.

3. Service Navigation: Many people at risk of homelessness are unaware of the services available to them. Outreach workers act as navigators, helping individuals access housing programs, mental health care, or financial assistance.

4. Culturally Competent Outreach: Outreach teams should

be trained to understand the unique needs of different communities, including immigrants, people of color, LGBTQ individuals, and those experiencing domestic violence, and tailor their approaches accordingly.

Targeted Outreach Programs

Targeted outreach programs focus on specific populations that are at higher risk of homelessness, such as:

- Low-Income Families: Providing resources and support to families facing financial hardship.

- Individuals with Mental Health or Substance Use Issues: Connecting these individuals to appropriate services before they reach a crisis point.

- Youth and Young Adults: Engaging with homeless youth or those at risk of homelessness, offering tailored support and resources.

Utilizing Data for Outreach

Data plays a crucial role in informing outreach strategies. By analyzing data sources such as HMIS, community needs assessments, and economic indicators, stakeholders can identify neighborhoods and populations most at risk of homelessness. Key strategies include:

1. Geospatial Analysis: Mapping areas with high eviction rates or low-income households can help target outreach efforts effectively.

2. Predictive Analytics: Using predictive models to identify individuals likely to experience homelessness can guide proactive interventions.

3. Community Surveys: Gathering qualitative data from community members can reveal barriers to accessing services and inform outreach approaches.

Data-Driven Approaches to Early Intervention and Outreach

The integration of data into early intervention and outreach strategies ensures that efforts are not only timely but also effective in reaching the right people and providing the appropriate services. By leveraging data, outreach and intervention programs can more precisely identify individuals at risk of homelessness and allocate resources efficiently. Data-driven outreach strategies help in targeting interventions before they reach a critical point of crisis.

Identifying At-Risk Individuals

Data-driven methods enable organizations to identify individuals and families at risk of homelessness before they lose their housing. By analyzing factors such as income fluctuations, eviction notices, and service utilization patterns, stakeholders can prioritize outreach efforts and allocate resources effectively.

Coordinated Entry Systems

Coordinated entry systems streamline access to housing and support services by providing a centralized point of contact for individuals experiencing or at risk of homelessness. These systems often rely on data to assess needs, prioritize individuals based on vulnerability, and connect them to appropriate resources quickly.

Program Evaluation and Adaptation

Monitoring and evaluating the effectiveness of early intervention and outreach programs are essential for continuous improvement. Key performance indicators (KPIs) may include:

- Success Rate of Prevention Efforts: The percentage of individuals who remain in their housing after receiving intervention services.

- Engagement Levels: The number of at-risk individuals who respond to outreach efforts and access services.

- Long-Term Outcomes: Tracking housing stability and overall well-being over time for individuals who received preventive assistance.

Predictive Analytics for Risk Identification

Predictive analytics is the process of using data to identify individuals who are at risk of homelessness. This involves analyzing patterns and risk factors from historical data to create predictive models that can forecast future homelessness. Key risk factors often include:

Recent job loss or financial instability

History of eviction or poor credit history

Substance abuse or mental health issues

Domestic violence or family conflict

Aging out of foster care or being recently incarcerated

By collecting and analyzing data from multiple sources, such as housing records, public health data, and criminal justice information, agencies can identify people at high risk of homelessness before they lose their homes. Data can also help predict which individuals might need immediate attention and which might benefit from long-term prevention efforts.

Case Studies of Successful Prevention Programs

Case Study 1: The Homebase Program in New York City

The Homebase program in New York City focuses on preventing homelessness by providing services and support to families at risk of eviction. Data analysis is used to identify high-risk neighborhoods and target outreach efforts effectively. The program has successfully reduced the number of families entering the shelter system by connecting them to financial assistance, legal support, and case management services.

Case Study 2: Los Angeles' Homelessness Prevention Initiative

Los Angeles has implemented a homelessness prevention initiative that combines data-driven outreach with community partnerships. By analyzing HMIS and eviction data, the city has identified areas with high rates of housing instability and deployed outreach teams to connect residents with resources. The initiative has demonstrated positive outcomes, reducing the number of households entering homelessness.

Challenges in Homelessness Prevention

1. Resource Limitations

Many outreach and prevention programs face resource constraints, limiting their ability to provide comprehensive support. Securing funding and resources is essential for expanding prevention efforts.

2. Stigma and Barriers to Access

Stigma surrounding homelessness and mental health can deter individuals from seeking help. Outreach strategies must prioritize building trust and reducing barriers to access services.

3. Data Integration

Integrating data from various sources can be challenging. Effective collaboration among service providers is necessary to ensure comprehensive data collection and analysis.

Engaging in homelessness prevention through early intervention and outreach is essential for breaking the cycle of homelessness. By prioritizing proactive measures and leveraging data-driven approaches, stakeholders can effectively identify at-risk individuals and provide the support needed to maintain housing stability. As Los Angeles continues to confront the challenges of homelessness, investing in prevention strategies will not only reduce the number of individuals experiencing homelessness but also enhance the overall health and resilience of the community. Through collaboration, innovation, and a commitment to early intervention, we can foster hope and create pathways to stability for those at risk.

Outreach

Outreach connects with individuals in public spaces, underserved areas, and community organizations. Peer outreach and community-based partnerships leverage local expertise and resources.

Examples:

1. The Salvation Army's Pathway of Hope Program utilizes peer outreach and community partnerships.

2. Seattle's Outreach Program employs trained outreach workers.

Effective Engagement Techniques

Building trust and rapport through empathetic relationships, trauma-informed care, and low-barrier entry minimize access obstacles. Technology enhances outreach efforts.

Examples:

1. Mobile apps, like Los Angeles' Homeless Outreach App, facilitate service access.

2. Online platforms, such as New York City's HOME portal, streamline service navigation.

Challenges and Considerations

Funding constraints, staff burnout, cultural and language barriers, and data collection complexities hinder effectiveness.

Best Practices for Implementation

Collaboration with community partners, ongoing staff training, cultural competence, and program evaluation ensure effective early intervention and outreach.

Emerging Trends

1. Integrated data systems enhance service coordination and evaluation.

2. Coordinated entry systems streamline service access.

3. Housing First approaches prioritize permanent housing.

Homelessness Prevention and Early Identification Tools

Several tools and platforms can help organizations track and analyze data to prevent homelessness:

Homeless Management Information Systems (HMIS): HMIS collects data on people seeking shelter, including demographics, housing status, and service usage. This data can help identify trends and pinpoint individuals who are at risk.

Screening and Risk Assessment Tools: Standardized risk assessment tools like the VI-SPDAT (Vulnerability Index - Service Prioritization Decision Assistance Tool) are used to assess the level of risk and prioritize resources for individuals based on their likelihood of experiencing homelessness.

3. Real-Time Data for Immediate Response

Real-time data allows outreach teams to respond to emerging needs as they arise. By integrating data collection tools with case management systems, agencies can track the status of individuals in real time and ensure they receive immediate assistance when needed. For example, if an outreach worker identifies someone in a vulnerable situation—such as an eviction notice or family dispute—they can immediately offer resources like financial assistance, legal support, or housing alternatives.

4. Geospatial Analysis for Targeting Outreach

Geospatial data, such as mapping tools, can assist in identifying areas with high concentrations of individuals at risk of homelessness. By using Geographic Information Systems (GIS), cities can identify neighborhoods with rising eviction rates or areas with high unemployment, and direct outreach efforts there. This geographic approach enables services to be more targeted and efficient, ensuring that resources are deployed where they are most needed.

5. Monitoring and Evaluation

Data also allows agencies to monitor the effectiveness of their outreach and early intervention efforts. By tracking the outcomes of individuals who have received assistance, agencies can assess which strategies are most successful at preventing homelessness. For example, analyzing whether financial assistance programs or housing subsidies result in sustained housing stability helps improve the design of future programs. Continuous data collection ensures that outreach efforts are continuously optimized.

Collaboration and Data Sharing

Early intervention and outreach efforts are most successful when they involve multiple agencies and organizations working together. For example, local governments, nonprofit organizations, and healthcare providers all play key roles in addressing the risk of homelessness.

Collaboration and data sharing between these entities are essential for providing coordinated services. When outreach workers, social services, and healthcare providers share data about clients (with consent), they can ensure that individuals receive comprehensive, continuous support. Cross-sector collaboration also facilitates information exchange about available resources, service gaps, and emerging trends, allowing communities to build a more effective and unified response to homelessness prevention.

Ethical Considerations in Early Intervention and Outreach

As with any data-driven effort, early intervention and outreach programs must prioritize the ethical handling of personal information. Key considerations include:

Confidentiality: Personal information, especially about vulnerabilities such as mental health or substance abuse, must be kept confidential and shared only with appropriate consent.

Informed Consent: Individuals must understand how their data will be used and be able to opt in or out of interventions.

Bias Reduction: Efforts should be made to ensure that data collection and intervention strategies are not biased toward certain groups, particularly marginalized populations.

Engaging with homelessness prevention through early intervention and outreach is critical. Addressing underlying issues, building trust, and providing supportive services prevent homelessness. By adopting best practices, addressing challenges, and integrating emerging trends, we can reduce homelessness.

Early intervention and outreach are crucial components of homelessness prevention. By identifying at-risk individuals and providing timely support, we can break the cycle of homelessness before it begins. Data-driven strategies enhance the effectiveness of these efforts, enabling communities to better target their resources, track progress, and optimize outcomes. As Los Angeles continues to refine its approach to homelessness, data-driven early intervention and outreach will remain essential to preventing future homelessness and supporting those in need.

CHAPTER 15: BUILDING A DATA-DRIVEN ECOSYSTEM: COLLABORATION AND PARTNERSHIPS

Addressing homelessness in Los Angeles requires a multifaceted approach that leverages the strengths of various stakeholders within a collaborative framework.

A data-driven ecosystem enhances the ability of organizations to work together, share resources, and utilize information effectively to tackle the complex challenges of homelessness. This chapter explores the importance of collaboration and partnerships in creating a robust, data-driven response to homelessness.

The Need for Collaboration

Complexities of Homelessness

Homelessness is not a singular issue but rather a confluence of various factors, including economic instability, mental health challenges, substance abuse, and systemic barriers.

Effectively addressing these interconnected issues necessitates collaboration among diverse stakeholders, including:

- Government Agencies: Local, state, and federal agencies play crucial roles in policy development, funding, and service provision.

- Nonprofit Organizations: Service providers that offer direct assistance, such as shelters, case management, and outreach programs.

- Healthcare Providers: Organizations that address the physical and mental health needs of homeless individuals.

- Community Groups: Local organizations that understand the unique challenges of their neighborhoods and can mobilize resources effectively.

- Academic Institutions: Researchers and scholars who can provide data analysis, evaluation, and insights to inform policy and practice.

Benefits of Collaborative Efforts

1. Comprehensive Solutions: Collaboration allows for a holistic approach, combining resources and expertise from multiple sectors to create comprehensive solutions that address the root causes of homelessness.

2. Resource Sharing: By pooling resources, organizations can maximize their impact, reduce duplication of services, and enhance service delivery to those in need.

3. Enhanced Data Collection and Sharing: Collaborative partnerships facilitate the sharing of data across organizations, leading to a more comprehensive understanding of homelessness and improved decision-making.

4. Increased Funding Opportunities: Collaborative initiatives can attract funding from government and philanthropic

sources by demonstrating a coordinated approach to addressing homelessness.

Building a Data-Driven Ecosystem

Establishing Data Standards

To create an effective data-driven ecosystem, stakeholders must establish common data standards and protocols. This includes:

- Defining Key Metrics: Agreeing on key performance indicators (KPIs) that will be used to measure success across different organizations.

- Data Collection Protocols: Developing standardized methods for collecting and reporting data to ensure consistency and reliability.

Data Sharing Agreements

Establishing data sharing agreements among organizations is essential for fostering collaboration. These agreements should outline:

- Data Ownership: Clarifying who owns the data and how it can be used.

- Privacy and Security: Implementing measures to protect sensitive information and ensure compliance with privacy regulations.

- Access and Use: Defining who has access to the data and how it can be utilized for analysis and decision-making.

Collaborative Data Platforms

Creating collaborative data platforms can facilitate the sharing and analysis of data across organizations. These platforms can provide:

- Centralized Data Access: A single source of truth where stakeholders can access up-to-date information on homelessness trends, service utilization, and outcomes.

- Visualization Tools: Tools that allow for the visualization of data trends, making it easier for stakeholders to identify patterns and inform decision-making.

- Analytical Capabilities: Features that enable organizations to conduct their analyses, leading to insights that can inform programming and resource allocation.

Case Studies of Successful Collaboration

Case Study 1: The Los Angeles Homeless Services Authority (LAHSA)

LAHSA serves as a coordinating body for homelessness services in Los Angeles County. By fostering collaboration among government agencies, nonprofit organizations, and community groups, LAHSA has created a comprehensive system for addressing homelessness. The agency utilizes a centralized data platform to track service usage, demographics, and outcomes, enabling stakeholders to make informed decisions and allocate resources effectively.

Case Study 2: The Homeless Outreach Program

A collaborative initiative involving local law enforcement, healthcare providers, and nonprofit organizations in Los Angeles focuses on outreach to individuals experiencing homelessness. By sharing data on service utilization and health outcomes, this partnership has improved the effectiveness of outreach efforts, connecting individuals to necessary resources and services.

Challenges to Collaboration

1. Organizational Silos

Many organizations operate in silos, focusing on their specific missions and objectives. Breaking down these silos requires a cultural shift and a commitment to collaboration among all stakeholders.

2. Funding Constraints

Limited funding can hinder collaborative efforts, as organizations may prioritize their own programs over joint initiatives. Advocating for funding that supports collaborative approaches is essential.

3. Data Privacy Concerns

Concerns about data privacy and security can impede data sharing efforts. Establishing clear protocols and fostering trust among organizations is crucial for overcoming these barriers.

Building a data-driven ecosystem through collaboration and partnerships is essential for effectively addressing homelessness in Los Angeles. By fostering cooperation among government agencies, nonprofit organizations, healthcare providers, and community groups, stakeholders can create comprehensive solutions that leverage shared resources and expertise.

As data sharing and collaboration continue to evolve, the potential for innovative approaches to homelessness will expand, ultimately leading to more effective interventions and improved outcomes for individuals experiencing homelessness. A commitment to collaboration not only enhances the capacity to respond to homelessness but also strengthens the fabric of the community, fostering resilience and hope for all residents.

CHAPTER 16: IMPLEMENTING DATA-DRIVEN SOLUTIONS: CASE STUDIES

The potential for data-driven solutions to address homelessness in Los Angeles is vast.

However, the real test lies in translating these theories and strategies into actionable, impactful programs. This chapter presents a series of case studies from both within Los Angeles and other regions that have successfully implemented data-driven solutions to tackle homelessness. These examples highlight the power of collaboration, data sharing, and technological innovation in improving outcomes for individuals experiencing homelessness.

Case Study 1: Los Angeles Homeless Services Authority (LAHSA) - Coordinated Entry System (CES)

One of the key initiatives in Los Angeles for addressing homelessness is the Coordinated Entry System (CES), developed by the Los Angeles Homeless Services Authority (LAHSA). This system streamlines the process of connecting people experiencing homelessness to housing and services through a centralized platform that integrates data from multiple service providers.

The Problem:

Before the CES was implemented, homelessness services in Los Angeles were fragmented, with clients often navigating a complex web of disconnected service providers. This inefficiency led to duplications in services, delays in housing placement, and unequal distribution of resources.

The Data-Driven Solution:

CES uses a unified data platform to collect and analyze information on the homeless population, including factors such as length of homelessness, health conditions, employment status, and housing needs. The system also incorporates a prioritization tool that assesses individuals' vulnerability based on these factors and allocates resources based on urgency.

Data from CES is shared across a network of more than 100 service providers, ensuring that individuals receive the appropriate care, whether it's emergency shelter, permanent housing, or supportive services. The centralized data allows for better case management, as service providers can track client progress and ensure continuous support.

Results:

Since its implementation, CES has led to improved efficiency in matching homeless individuals with housing and services. The system's ability to prioritize those most at risk of chronic homelessness has led to more successful housing placements and fewer people cycling through temporary shelters. Additionally, the streamlined data-sharing has improved collaboration among service providers, ensuring no one falls through the cracks.

Case Study 2: The Home For Good Initiative - Los Angeles

Another notable case in Los Angeles is the Home For Good

Initiative, a collaboration of nonprofits, local government, and business leaders aimed at reducing homelessness through coordinated services and data-driven strategies. This initiative leverages a combination of funding, data, and partnership to offer housing and supportive services to the most vulnerable individuals.

The Problem:

The Home For Good Initiative was launched in response to rising homelessness rates in Los Angeles, especially among families, veterans, and chronically homeless individuals. A significant barrier to progress was the inability to connect these groups with timely services due to a lack of data sharing and coordination across sectors.

The Data-Driven Solution:

Home For Good has created a data-driven ecosystem by bringing together public and private stakeholders to pool resources.

The Housing First approach is based on the principle that providing stable housing is the first step toward addressing other challenges faced by individuals experiencing homelessness, such as mental health issues, substance abuse, and employment. In Utah, this model has been successfully implemented with the help of data-driven strategies that focus on housing stability as the starting point for long-term recovery.

The Problem:

Utah faced a significant challenge with chronic homelessness, particularly among individuals with severe mental health issues and substance abuse problems. The traditional approach of offering temporary shelter followed by access to services didn't address the root causes of homelessness and often resulted in individuals cycling in and out of shelters, without making long-term progress.

The Data-Driven Solution:

Utah's Housing First approach combined stable housing with wraparound services such as mental health care, substance abuse treatment, and job training. The state used data to identify those most in need and ensure that housing resources were allocated efficiently. Predictive modeling and risk assessments were used to identify individuals at high risk of long-term homelessness, ensuring that they were prioritized for immediate placement into housing.

Additionally, Utah collected and analyzed data on the effectiveness of the Housing First model, tracking outcomes such as the length of time individuals stayed housed, improvements in health and employment, and reductions in emergency healthcare and criminal justice system usage. This data-driven feedback loop helped fine-tune services and optimize resource allocation.

Results:

Utah's Housing First initiative led to a dramatic reduction in chronic homelessness. Since its implementation, the state has decreased its chronic homeless population by over 90%, becoming a model for other states and cities. The combination of stable housing and supportive services led to improved health, increased employment, and a significantly lower reliance on emergency services. The success of Utah's approach highlights the power of data-driven resource allocation in achieving long-term solutions for homelessness.

Case Study 4: The 100,000 Homes Campaign - National

The 100,000 Homes Campaign was a national initiative aimed at housing 100,000 homeless Americans, with a specific focus on veterans, chronically homeless individuals, and families. This campaign used data-driven methods to identify individuals who were most in need of housing and connect them with available resources.

The Problem:

Despite the availability of housing programs, there was often a mismatch between the needs of homeless individuals and the resources available. Many programs failed to prioritize those with the highest needs or allocate resources in the most efficient way. Additionally, service providers lacked real-time data on the availability of housing and the specific needs of homeless individuals.

The Data-Driven Solution:

The 100,000 Homes Campaign employed a "registry week" approach, where communities across the nation conducted surveys to collect data on the most vulnerable individuals experiencing homelessness. This data included information on medical conditions, mental health status, and other factors that contributed to homelessness. The campaign also used a centralized Homeless Management Information System (HMIS) to match individuals with available housing.

Through predictive analytics and data-driven outreach, the campaign was able to identify those at the greatest risk of dying on the streets or remaining homeless for an extended period. The data helped service providers tailor their interventions and allocate resources effectively.

Results:

The 100,000 Homes Campaign was a resounding success, housing over 105,000 individuals across the United States, including a significant number of veterans and chronically homeless individuals. The use of data to prioritize housing placements based on vulnerability and need ensured that the most at-risk individuals were given top priority. This approach showed that when data is used to guide service delivery, resources are allocated more effectively, and the success of homelessness interventions is significantly improved.

Case Study 5: The Homeless Outreach Program - San Francisco

San Francisco's Homeless Outreach Program is a prime example of how targeted, data-driven outreach efforts can effectively reduce homelessness in urban areas. The program focuses on engaging with individuals living on the streets and connecting them to housing, healthcare, and other critical services.

The Problem:

San Francisco faced a growing homeless population, with many individuals living in public spaces and struggling to access services. Outreach workers were often unable to identify who was in most need or where to focus their efforts, leading to inefficient use of resources and missed opportunities to engage individuals in housing and support services.

The Data-Driven Solution:

The Homeless Outreach Program uses real-time data collection through mobile applications and databases to track interactions with individuals experiencing homelessness. Outreach workers input data on individuals they encounter, including their specific needs, health conditions, and housing preferences. This data is shared across a network of service providers, allowing for coordinated efforts and timely interventions.

The program also leverages geospatial analysis to map areas with the highest concentrations of homeless individuals, enabling outreach workers to target areas where services are most needed. Predictive analytics are used to identify individuals at high risk of long-term homelessness, ensuring that they are prioritized for outreach and housing placement.

Results:

San Francisco's Homeless Outreach Program has led to a significant increase in the number of individuals engaged with housing and services. The data-driven approach has improved outreach efficiency, ensuring that resources are allocated to the individuals who need them most. Additionally, the program has helped reduce the number of individuals living on the streets by connecting them with long-term housing solutions and support services.

These case studies demonstrate the power of data-driven solutions in addressing homelessness. Whether through centralized data platforms, predictive analytics, geospatial mapping, or targeted outreach, data has proven to be an essential tool in the fight against homelessness. By leveraging data to prioritize resources, improve service delivery, and coordinate efforts across agencies, communities can make significant strides toward ending homelessness and improving the lives of those affected.

As we move forward, it is clear that the success of these programs relies not only on data but also on collaboration, commitment, and a shared vision for a future where homelessness is a problem of the past. Data is not just a tool; it is a key to unlocking sustainable, compassionate solutions for homelessness.

CHAPTER 17: EVALUATING EFFECTIVENESS: DATA-DRIVEN PERFORMANCE METRICS

One of the fundamental challenges in addressing homelessness is ensuring that the resources, interventions, and policies in place are effective in reducing the number of individuals experiencing homelessness.

W hile there are numerous initiatives and programs aimed at this issue, measuring their effectiveness is crucial to achieving long-term success. Data-driven performance metrics offer a powerful framework to assess the impact of homelessness programs, identify areas for improvement, and optimize resource allocation.

The Importance of Performance Metrics in Homelessness Solutions

To truly address homelessness, it is essential to know what is working and what is not. Performance metrics help answer this question by providing tangible, measurable data that can track progress over time. Without these metrics, there is no way to quantify success, make informed decisions, or demonstrate accountability to stakeholders and the public.

In the context of homelessness, performance metrics are not limited to the number of individuals housed or the cost of services provided. They also encompass a range of indicators that can show how well systems are supporting individuals, helping them regain stability, and preventing future homelessness. These metrics provide insight into the effectiveness of both the Housing First model and Rapid Rehousing approaches, which prioritize permanent housing as the first step to overcoming homelessness.

Key Performance Metrics for Homelessness Programs

The key to effective performance evaluation lies in selecting the right metrics that align with program goals and desired outcomes. Here are several essential performance metrics that should be used to evaluate homelessness interventions:

Housing Retention Rates

One of the most direct measures of success is whether individuals who are housed remain in stable housing. High retention rates indicate that programs are successful in offering the support needed for long-term stability, whereas low retention rates might suggest that additional services (such as mental health counseling, substance abuse treatment, or employment support) are needed to ensure success.

How to measure: Track the number of individuals placed in housing and the percentage who remain housed after 6 months, 12 months, and beyond. This metric is also known as housing stability.

Time to Housing

The time to housing metric measures how quickly individuals experiencing homelessness are placed into permanent housing.

The quicker people are housed, the more likely they are to experience stability and avoid the negative health, social, and economic consequences associated with long-term homelessness.

How to measure: Calculate the average number of days it takes to place an individual from initial contact with a service provider to permanent housing placement. A shorter duration is preferable.

Exits to Permanent Housing

The rate of exits to permanent housing is a key indicator of program effectiveness. This metric tracks the number of individuals who transition from homelessness to stable, permanent housing, whether through Rapid Rehousing or the Housing First approach.

How to measure: Track the percentage of individuals who exit homelessness and are permanently housed, with a particular focus on those who do not return to homelessness within a specified period, such as one year.

Client Health and Well-Being

Programs that address homelessness should also aim to improve the physical and mental health of individuals. Improved health and well-being are indicators that the system is offering holistic support that addresses the underlying causes of homelessness, such as mental health issues and substance abuse.

How to measure: Track health outcomes, including improvements in mental health, substance abuse treatment completion, and reduced reliance on emergency health services. Surveys, case management reports, and follow-up health assessments can be used to collect this data.

Employment and Income Outcomes

One of the main challenges for individuals exiting homelessness is the ability to secure stable employment and sustainable income. Employment and income outcomes provide a measure of the success of homelessness interventions in helping individuals regain self-sufficiency.

How to measure: Track the employment rate of individuals who have exited homelessness and the number of individuals who have achieved stable income. This can include jobs, benefits, or other forms of economic support.

Reduction in Emergency Service Usage

One of the most effective ways to measure the success of homelessness programs is by tracking how often individuals rely on costly emergency services, such as emergency room visits, police interventions, or jail time. Effective programs should reduce the need for these emergency interventions by stabilizing individuals in housing and providing supportive services.

How to measure: Monitor the frequency of emergency service utilization by program participants before and after they are housed. This data can be collected in partnership with hospitals, police departments, and other service providers.

Cost-Effectiveness

While not always an immediate priority, evaluating the cost-effectiveness of homelessness interventions is critical in demonstrating the long-term sustainability of programs. Programs that are cost-effective provide more value per dollar spent and are more likely to attract long-term funding and support.

How to measure: Compare the cost of housing individuals and providing services (such as healthcare, case management, and housing support) to the savings generated through reduced reliance on emergency services, public healthcare, and criminal justice involvement.

Using Data to Optimize Performance

The collection of performance metrics is only the first step in improving homelessness programs. The real value comes from using this data to make informed decisions and optimize program delivery. Here's how data can be leveraged:

Data-Driven Adjustments to Program Design

Analyzing performance metrics allows service providers to identify patterns and trends in outcomes, helping them understand what's working and what's not. For example, if housing retention rates are low in a specific region, it may signal that additional wraparound services, such as mental health care, are needed in that area. Conversely, if employment outcomes are poor, it may suggest that job training programs need to be improved.

Feedback Loops for Continuous Improvement

Programs that collect data on performance should have built-in feedback loops that allow for continuous improvement. By regularly reviewing data, stakeholders can adjust their strategies, resources, and outreach efforts based on real-time information. For example, if a specific intervention is found to be ineffective, adjustments can be made to make it more successful.

Cross-Collaboration and Data Sharing

The effectiveness of homelessness interventions can be significantly enhanced when data is shared between different service providers, such as housing authorities, healthcare providers, and social service organizations. Collaboration allows for a holistic view of each individual's situation and the ability to track their progress across multiple systems.

Predictive Analytics for Early Intervention

By using historical data and predictive models, service providers can anticipate which individuals are most at risk of falling into homelessness. Early intervention, guided by predictive analytics, can help prevent homelessness before it occurs by targeting resources at the individuals who need them most.

Challenges in Performance Evaluation

While data-driven performance metrics are invaluable, there are also several challenges that must be addressed:

Data Collection Issues: Inconsistent or incomplete data can lead to inaccurate assessments. Ensuring that all relevant data is collected and standardized across organizations is essential.

Privacy Concerns: Collecting data on homeless individuals requires sensitivity and respect for privacy. Ensuring that data is securely stored and anonymized is critical to protect individuals' rights.

Resource Constraints: Evaluating homelessness interventions using data can be resource-intensive. Organizations may face challenges in collecting and analyzing large volumes of data while still providing essential services.

Data-driven performance metrics are essential for evaluating the effectiveness of homelessness programs. By tracking key indicators, such as housing retention, health outcomes, employment, and cost-effectiveness, organizations can assess the success of their interventions and optimize service delivery. With a commitment to using data for continuous improvement, we can move closer to ending homelessness, ensuring that resources are used efficiently and that every individual receives the support they need to achieve long-term stability.

Chapter 19: Emerging Trends and Technologies: Opportunities and Challenges

As we continue to innovate and develop solutions to address homelessness, emerging trends and technologies are playing an increasingly vital role in shaping the landscape of homelessness prevention, intervention, and service delivery. These trends not only offer new opportunities for improving the efficiency and effectiveness of programs but also present a set of challenges that must be navigated carefully.

In this chapter, we will explore the latest technological advancements and trends influencing the fight against homelessness, the potential they offer for solving this complex issue, and the challenges they introduce. From advancements in artificial intelligence (AI) and data analytics to the application of blockchain and digital platforms, these technologies have the potential to reshape homelessness interventions, but they must be harnessed thoughtfully to ensure their benefits are realized for all populations.

Emerging Trends and Technologies Shaping Homelessness Solutions

Artificial Intelligence and Machine Learning

Artificial intelligence (AI) and machine learning (ML) are increasingly being used to analyze large sets of data, predict trends, and identify the most effective interventions for individuals experiencing homelessness. These technologies offer powerful tools for creating predictive models that can help communities anticipate homelessness risk and intervene proactively.

For instance, machine learning algorithms can process historical data on individuals' pathways into homelessness and predict who is at risk, enabling service providers to offer targeted interventions before a person loses their home. Additionally, AI-powered chatbots and virtual assistants can assist case managers by providing real-time information and support for clients.

Opportunity: Improved prediction of homelessness risk, tailored service delivery, and automated support for case management.

Challenge: Ensuring that AI models are accurate and free from biases, especially in communities with complex socio-economic dynamics.

Blockchain for Data Security and Transparency

Blockchain technology has the potential to revolutionize how data is managed and shared across different agencies and organizations working to address homelessness. With its decentralized nature, blockchain can ensure that personal data is securely stored and shared, providing greater privacy and control to individuals experiencing homelessness while enabling seamless coordination among service providers.

Blockchain can also be used for the management of resources, such as housing vouchers, social benefits, or financial aid, ensuring transparency, reducing fraud, and providing a clear, auditable record of transactions.

Opportunity: Increased data security, transparency in resource distribution, and streamlined coordination between agencies.

Challenge: Implementing blockchain in the homelessness sector requires overcoming technical barriers, including integration with existing systems and ensuring accessibility for all users.

Mobile and Digital Platforms for Service Access

Mobile apps and digital platforms are transforming how individuals access services and support. For those experiencing homelessness, access to services can often be challenging due to mobility issues, lack of stable internet, or limited physical infrastructure. Digital platforms can bridge this gap by offering mobile-based solutions that allow individuals to connect with resources such as shelters, food programs, healthcare services, and job opportunities.

Apps can also provide real-time information, such as available shelter space or transportation options, and enable direct communication with case managers or service providers.

Opportunity: Enhanced access to services and information, improved communication between service providers and clients, and increased self-sufficiency for individuals experiencing homelessness.

Challenge: Digital access may be limited among certain homeless populations, and there may be challenges in ensuring the security and privacy of sensitive personal information.

Data Analytics and Real-Time Monitoring

Data analytics is playing an essential role in monitoring homelessness trends and optimizing service delivery. By utilizing real-time data, service providers can track the availability of resources, monitor the effectiveness of interventions, and quickly adjust strategies based on emerging patterns. Real-time monitoring allows for better coordination, ensuring that services are available when and where they are needed most.

For instance, data dashboards can track shelter occupancy rates, medical services utilization, and the status of housing placements, helping service providers allocate resources dynamically and prevent bottlenecks in service delivery.

Opportunity: Enhanced decision-making, real-time resource allocation, and better understanding of homelessness trends.

Challenge: The need for accurate, up-to-date data and the challenge of integrating data from multiple sources without overwhelming caseworkers or service providers.

Virtual and Augmented Reality for Housing Solutions

Virtual reality (VR) and augmented reality (AR) technologies are beginning to be applied in innovative ways to address housing issues. VR can be used to offer virtual tours of housing options, enabling individuals experiencing homelessness to view potential homes remotely. AR applications can assist in visualizing potential modifications to housing spaces, such as for accessibility or safety purposes.

Moreover, VR can serve as a training tool for both service providers and individuals experiencing homelessness, helping case managers practice various scenarios and improve their skills in interacting with clients.

Opportunity: Enhanced access to housing options, improved training for service providers, and the ability to remotely view housing opportunities.

Challenge: The cost and availability of VR/AR technology, as well as the need to ensure that these technologies are user-friendly and accessible to individuals who may have limited technological literacy.

Smart Cities and Internet of Things (IoT)

The concept of smart cities is being explored as a way to create more connected, sustainable, and efficient urban environments. Smart city technologies, powered by the Internet of Things

(IoT), can be used to improve homelessness interventions. For instance, IoT sensors can monitor shelter conditions, track air quality, and optimize the delivery of resources like food or medical supplies.

Additionally, smart infrastructure, such as public Wi-Fi and smart transportation systems, can provide greater access to essential services for individuals experiencing homelessness, helping them navigate the challenges of urban living.

Opportunity: Greater integration of technology in urban spaces, improved resource delivery, and enhanced living conditions for individuals experiencing homelessness.

Challenge: Ensuring that the benefits of smart city technologies are equitably distributed and accessible to all members of the community, including those experiencing homelessness.

Challenges in Leveraging Emerging Technologies

While the opportunities presented by emerging technologies are vast, there are several challenges that need to be addressed to ensure that these innovations effectively serve homeless populations. These challenges include:

Equity and Access

Not all individuals experiencing homelessness have access to smartphones, computers, or reliable internet. Digital divides, especially in low-income and homeless communities, can limit the effectiveness of technology-driven solutions. Ensuring that technological interventions are accessible to all individuals— regardless of their access to devices or internet connectivity—is essential to the success of these innovations.

Data Privacy and Security

As more personal data is collected through digital platforms, mobile apps, and AI-powered tools, safeguarding the privacy and security of this data becomes critical. Homeless individuals are often vulnerable to identity theft, fraud, or misuse of their personal information. Implementing strong data protection measures is vital to maintaining trust and ensuring that individuals' sensitive information is kept secure.

Cost and Implementation Barriers

The implementation of emerging technologies can be costly, both in terms of initial investments and ongoing maintenance. For cash-strapped service providers and government agencies, funding and resources may not be sufficient to incorporate high-tech solutions at scale. Moreover, technological solutions require training and support for both service providers and homeless individuals, which can add to the cost and complexity of implementation.

Integration with Existing Systems

Integrating new technologies into existing systems and workflows can be complex. Homelessness service providers often rely on outdated systems or disconnected databases, making it difficult to introduce and maintain new technological tools. Overcoming technical barriers to integration is a key challenge when scaling technological solutions in the homelessness sector.

Ethical Considerations

The use of AI, blockchain, and other technologies raises important ethical considerations. For example, algorithmic bias in AI models could exacerbate inequalities or unfairly target certain populations. Similarly, blockchain applications need to ensure that the use of personal data does not inadvertently stigmatize or harm individuals experiencing homelessness.

Opportunities for Innovation and Future Trends

As technology continues to evolve, there are several emerging trends that hold promise for addressing homelessness in new and innovative ways. These trends include:

AI-Powered Predictive Tools for early intervention and

personalized services.

5G Networks and Connectivity to ensure reliable access to digital platforms and resources for homeless individuals.

Smart Housing Solutions such as modular, affordable housing built with technology-driven efficiency.

Wearable Technologies that track health and safety, alerting service providers in real-time to potential issues.

As we move forward, it is important to remain open to new ideas and solutions. By embracing these emerging technologies thoughtfully and ethically, we can create a future where homelessness is not only addressed more effectively but ultimately eradicated.

Emerging trends and technologies offer exciting opportunities for improving how we address homelessness, making interventions more targeted, efficient, and accessible. However, with these opportunities come significant challenges that must be navigated carefully. By addressing equity, privacy, and integration concerns, and ensuring that technological solutions are used ethically and sustainably, we can harness the power of innovation to drive meaningful change in the fight against homelessness. As we continue to explore the potential of these technologies, we must keep in mind that the ultimate goal is to create a system that works for everyone, leaving no one behind.

Chapter 20: Policy and Advocacy: Using Data to Inform Decision-Making

In the effort to address homelessness, data is not just a tool for service delivery but also a powerful instrument for shaping policy, driving advocacy, and influencing decision-making at local, state, and national levels. By leveraging data effectively, we can identify gaps in existing systems, demonstrate the impact of homelessness interventions, and advocate for the resources and reforms necessary to create lasting solutions. This chapter explores the role of data in informing policy and advocacy, providing insights on how data can drive systemic change and improve outcomes for individuals experiencing homelessness.

The Role of Data in Shaping Homelessness Policy

Homelessness is a complex, multifaceted issue that requires comprehensive and coordinated responses across various sectors, including housing, healthcare, social services, and employment. Data serves as the foundation for designing, implementing, and evaluating policies aimed at reducing homelessness. Policymakers need accurate, timely, and reliable data to understand the scope of the problem, identify trends, and assess the effectiveness of existing programs.

Informing Policy Development

Data can inform the development of policies that address the root causes of homelessness, including lack of affordable housing, economic inequality, mental health issues, and substance abuse. By analyzing data from various sources, policymakers can craft targeted policies that respond to the unique needs of different homeless populations, such as families, veterans, individuals with disabilities, or those experiencing chronic homelessness.

For example, data on the availability of affordable housing can be used to advocate for zoning changes or increased funding for housing subsidies. Data on the prevalence of mental health or substance use disorders among the homeless population can lead to the creation of more comprehensive healthcare services tailored to these individuals' needs.

Opportunity: Data enables policymakers to design evidence-based policies that are more likely to achieve positive outcomes and effectively address homelessness.

Challenge: Data must be comprehensive and accurately reflect the diverse factors contributing to homelessness, or policies may fail to address underlying causes.

Assessing Program Effectiveness

Data can also be used to evaluate the effectiveness of homelessness programs and services. By tracking key performance indicators (KPIs), such as shelter occupancy rates, housing retention, employment outcomes, and health improvements, policymakers can assess which programs are working and which need to be adjusted or discontinued.

For example, if a rapid rehousing program shows high success rates in quickly placing individuals into stable housing, data can be used to advocate for the expansion of that program. Conversely, if a program is found to be ineffective, data can highlight areas for improvement or inform the decision to reallocate resources to more successful initiatives.

Opportunity: Data-driven evaluations allow for continuous improvement of homelessness programs, ensuring that resources are used effectively to achieve the best possible outcomes.

Challenge: Data collection and analysis must be consistent and standardized across programs to ensure comparability and accurate assessments.

Advocating for Policy Reform

Advocacy is a critical part of driving systemic change, and data plays a central role in this process. By presenting compelling data that highlights the severity of homelessness and the effectiveness of certain interventions, advocates can persuade policymakers, funders, and the public to prioritize homelessness as a key issue.

Data can be used to demonstrate the economic costs of homelessness, such as the strain on emergency services, healthcare, and criminal justice systems, as well as the potential savings from prevention and intervention programs. For instance, studies showing that every dollar spent on homelessness prevention saves multiple dollars in healthcare and criminal justice costs can be a powerful tool for advocating for increased funding for prevention and rapid rehousing programs.

Opportunity: Data can provide the evidence needed to build a strong case for policy reforms and secure funding for homelessness programs.

Challenge: Advocacy efforts must ensure that data is presented in an accessible and persuasive manner to reach diverse audiences, including policymakers, donors, and the public.

Types of Data Essential for Policy and Advocacy

Different types of data are necessary for informing effective policies and advocacy efforts. Below are some key types of data that play a significant role in addressing homelessness:

Demographic Data

Understanding the demographics of the homeless population is essential for designing targeted interventions. Data on factors such as age, gender, race, family composition, and immigration status can help policymakers develop policies that address the specific needs of different subpopulations.

Example: A growing number of homeless youth or families with children may prompt the development of more family-centered housing programs and child welfare services.

Service Utilization Data

Data on the use of homelessness services—such as shelters, food programs, healthcare, and employment assistance—helps policymakers identify gaps in service delivery and allocate resources more efficiently. Tracking how frequently services are accessed and by whom can reveal patterns and guide decisions on where to expand or improve services.

Example: If a region experiences high demand for shelters but a low success rate in housing placements, policymakers may prioritize expanding rapid rehousing initiatives or increasing the availability of long-term supportive housing.

Housing Data

Housing data, including the availability of affordable housing, vacancy rates, rent levels, and the number of housing subsidies or vouchers issued, is critical for informing policies aimed at increasing access to stable housing for homeless individuals.

Example: Data on the shortage of affordable housing can be used to advocate for new affordable housing projects, zoning reforms, or the expansion of housing voucher programs.

Health and Social Services Data

Homelessness is often intertwined with health and social issues, such as mental health conditions, substance use disorders, chronic illness, and trauma. Data on health and social services utilization can help identify trends and gaps in service provision, enabling policymakers to create integrated health and housing solutions.

Example: Data on the prevalence of mental health disorders among homeless populations can support the creation of specialized housing-first programs that provide both stable housing and mental health services.

Economic and Employment Data

Employment and income data is key for understanding the economic factors contributing to homelessness. Data on unemployment rates, wages, and income inequality can inform policies aimed at addressing the economic drivers of homelessness.

Example: If data shows a high correlation between unemployment and homelessness in certain regions, policies may focus on job training programs, wage increases, and employment opportunities for homeless individuals.

Using Data to Influence Policy at Various Levels

Data-driven advocacy is essential for influencing policy decisions at the local, state, and national levels. Here's how data can be used to affect policy at each level:

Local Level

At the local level, data is often used to advocate for changes in zoning laws, the development of affordable housing, or the allocation of funding for homelessness programs. Local government officials rely on data to identify the most pressing needs in their communities and prioritize interventions accordingly.

Example: In cities with high rates of homelessness, data showing the effectiveness of permanent supportive housing programs can be used to encourage local governments to prioritize funding for these initiatives.

State Level

State governments often control funding for homelessness programs and have the ability to enact policies that affect the broader housing and social welfare systems. Data on statewide homelessness trends, such as regional disparities in housing availability or the effectiveness of eviction prevention programs, can be used to advocate for state-level reforms.

Example: Data showing the success of a particular state's eviction prevention program can be used to advocate for the replication of that model in other regions of the state.

National Level

At the national level, data is used to inform large-scale policy decisions, including federal funding allocations, the creation of national initiatives to combat homelessness, and the development of comprehensive homelessness prevention strategies.

Example: Data on the high costs of chronic homelessness, including emergency healthcare and justice system involvement, can be used to argue for greater federal investment in homelessness prevention and housing-first programs.

Challenges in Using Data for Policy and Advocacy

While data is an invaluable tool for shaping policy and advocacy efforts, several challenges can hinder its effectiveness:

Data Gaps and Quality Issues

Incomplete or poor-quality data can undermine advocacy efforts. Inaccurate or missing data on homelessness trends, service utilization, or program effectiveness can lead to misguided policy decisions.

Privacy and Confidentiality Concerns

Data collection and use must balance the need for comprehensive information with the protection of individuals' privacy. Ensuring that data is collected and shared responsibly is essential to maintaining public trust and safeguarding vulnerable populations.

Political and Bureaucratic Barriers

In some cases, political or bureaucratic obstacles can delay or prevent the use of data to inform policy decisions. Decision-makers may be resistant to adopting data-driven solutions if they conflict with existing policies or political priorities.

Best Practices for Data-Driven Advocacy and Policy Implementation

To maximize the impact of data in advocacy and policymaking, organizations and advocates must adhere to best practices that enhance the reliability, transparency, and accessibility of data. The following strategies ensure that data serves its full potential in the fight against homelessness:

1: Data Standardization and Consistency

One of the critical aspects of successful data-driven policymaking is the standardization of data. This ensures that data collected from different sources, regions, or organizations can be compared and analyzed effectively. Standardized data allows policymakers to make informed decisions based on reliable, apples-to-apples comparisons.

Actionable Strategy: Advocate for the adoption of common data standards across different agencies, service providers, and jurisdictions involved in homelessness services. This could include using standardized case management tools, consistent definitions for homelessness, and uniform reporting formats.

Impact: Consistent data helps ensure that interventions are aligned with regional and national standards, facilitating collaboration and scaling up successful approaches.

2: Data Transparency and Accessibility

For data to be impactful in policy and advocacy, it must be easily accessible to those who need it. This means not only making data available to policymakers but also to community organizations, the public, and people directly affected by homelessness.

Actionable Strategy: Develop online platforms and dashboards where homelessness data can be shared in a user-friendly format. These tools can help communities, researchers, and policymakers access real-time information and analyze trends in homelessness.

Impact: When data is accessible and transparent, it fosters accountability, allows for public scrutiny, and encourages informed decision-making at all levels.

3: Collaboration Across Sectors

The complex nature of homelessness means that addressing it requires collaboration across multiple sectors, including housing, healthcare, education, and employment. Effective data-driven policy solutions must be the product of cross-sector collaboration.

Actionable Strategy: Create formal partnerships between government agencies, nonprofit organizations, businesses, healthcare providers, and academic institutions. These partnerships can help pool data from diverse sources, ensuring a comprehensive understanding of homelessness.

Impact: Collaborating across sectors allows for holistic interventions that address the multiple facets of homelessness, ensuring that solutions are comprehensive and sustainable.

4: Engaging Affected Communities in Data Collection

While it is crucial to use data to inform decision-making, it is equally important to involve the homeless population in the data collection process. Engaging individuals with lived experience ensures that the data collected is reflective of their needs and experiences, rather than being driven solely by external perspectives.

Actionable Strategy: Incorporate feedback from individuals experiencing homelessness in the design and implementation of data collection efforts. Use participatory research methods, such as interviews, focus groups, and community surveys, to gather insights directly from those affected.

Impact: This approach ensures that the data is more accurate and relevant to the lived experiences of homeless individuals, which in turn can help create policies that are more empathetic and effective.

5: Real-Time Data for Rapid Response

In order to be most effective, data must not only be accurate but also timely. Having real-time data allows for quick responses to emerging homelessness trends or immediate needs.

Actionable Strategy: Implement systems that allow for the real-time collection and analysis of data, such as mobile applications for service providers, GPS tracking for outreach teams, or automated data-entry systems.

Impact: Real-time data allows for quicker adjustments to service delivery, more immediate interventions, and better crisis management.

Data is a powerful tool in the fight against homelessness. It provides the evidence necessary to inform policymaking, evaluate the effectiveness of interventions, and advocate for the resources needed to address the issue. By leveraging data effectively

Harnessing Big Data and Technology in Homelessness Policy

As technology continues to evolve, new opportunities to leverage big data and advanced analytics for homelessness policy emerge. From predictive modeling to geospatial analysis, the possibilities for using data to tackle homelessness are expanding rapidly. Below are some cutting-edge technologies that can be used to influence policy and advocacy efforts:

Predictive Analytics and Forecasting

Predictive analytics uses historical data and advanced algorithms to forecast future trends and behaviors. By analyzing factors such as housing market trends, economic conditions, and previous service usage, predictive models can identify individuals or populations at risk of homelessness before they become homeless.

Actionable Strategy: Use predictive models to identify high-risk individuals and offer proactive interventions, such as eviction prevention programs, early case management, or financial assistance.

Impact: Predictive analytics can help prevent homelessness before it occurs, enabling policymakers and service providers to address issues early and avoid costly interventions later.

Geospatial Data and Mapping

Geospatial data analysis involves using geographic information systems (GIS) to visualize and analyze spatial data. This can be particularly useful for identifying patterns in homelessness and service access, helping policymakers determine the best locations for resource allocation.

Actionable Strategy: Utilize GIS tools to map areas with high concentrations of homelessness, identify gaps in services, and prioritize areas for new housing or shelter development.

Impact: GIS mapping allows for smarter resource allocation, ensuring that interventions are implemented in areas where they are most needed.

Artificial Intelligence and Machine Learning

Artificial intelligence (AI) and machine learning (ML) can

be used to automate data processing, identify patterns, and predict outcomes. These technologies can improve the speed and accuracy of data analysis, helping policymakers make data-driven decisions more quickly.

Actionable Strategy: Implement AI-powered data analysis tools that can sift through large datasets, identify trends, and predict the success of different homelessness interventions.

Impact: By automating routine data analysis, AI and ML can help policymakers and advocates make faster, more informed decisions about how to address homelessness.

Challenges and Ethical Considerations in Data-Driven Policy

Despite the immense potential of data to inform policy and advocacy, there are several challenges and ethical considerations that must be addressed to ensure that data is used responsibly:

Privacy and Confidentiality

Homeless individuals are often vulnerable to exploitation and discrimination. Ensuring the privacy and confidentiality of the data collected is critical to building trust with the community and protecting individuals' rights.

Solution: Implement strict data protection protocols, such as encryption, de-identification, and secure data storage, to ensure that sensitive information is protected.

Bias and Data Quality

Data can sometimes reflect biases in its collection or interpretation, which can lead to skewed outcomes. It is essential to ensure that data is representative and unbiased, especially when it comes to marginalized populations such as homeless individuals.

Solution: Engage with communities affected by homelessness to ensure that their perspectives and experiences are accurately represented in data collection efforts. Regularly audit data for biases and gaps.

Accessibility and Equity

There is a risk that data-driven policies may disproportionately benefit those who already have access to resources, leaving out marginalized communities. Ensuring that data policies are equitable and inclusive is essential for addressing the root causes of homelessness.

Solution: Advocate for policies that ensure data collection and analysis efforts are inclusive, reflecting the diversity of experiences within the homeless population.

Data as a Catalyst for Change

Data has the power to transform how we address homelessness, from shaping policies and advocacy efforts to optimizing service delivery and identifying new opportunities for intervention. By employing data strategically and ethically, we can create more effective, sustainable solutions to homelessness, improving outcomes for individuals and communities.

To achieve this, we must prioritize data collection, standardization, and accessibility, invest in new technologies, and foster collaboration across sectors. With these efforts, data can be harnessed to create systemic change, drive policy reforms, and ultimately end homelessness. The future of homelessness prevention and intervention lies in data-driven strategies that are responsive, adaptive, and inclusive, ensuring that everyone has a pathway to stability and self-sufficiency

Chapter 22: Community Engagement: Empowering Local Communities

Addressing homelessness in Los Angeles is not just a matter of policy and resources but also requires the active involvement and empowerment of local communities. Community engagement is essential for creating tailored, effective solutions that address the specific needs of each neighborhood. This chapter explores strategies for involving community members, fostering local leadership, and leveraging community resources to make a lasting impact.

The Importance of Community Engagement in Tackling Homelessness

Community engagement fosters a sense of shared responsibility and partnership in addressing homelessness. When communities actively participate in creating solutions, the resulting initiatives are more likely to reflect local values, gain public support, and lead to sustainable outcomes.

Benefits of Community Engagement:

Building Trust: Establishing trust between service providers, policymakers, and community members.

Local Insights: Leveraging the unique perspectives and experiences of residents to create relevant solutions.

Increased Accountability: Encouraging transparency and community oversight, which strengthens accountability.

Understanding Community Needs Through Participatory Research

Engaging communities in the research process helps ensure that data accurately reflects their needs and priorities. Participatory research methods, such as community surveys, focus groups, and town hall meetings, allow community members to share their insights, shaping the direction of homelessness initiatives.

Participatory Research Techniques:

Community Surveys: Collect data on local experiences, concerns, and ideas.

Focus Groups: Facilitate discussions with diverse community members to understand a broad range of perspectives.

Community-Based Advisory Boards: Establish boards with local representatives who can provide ongoing input.

Empowering Local Leaders as Change Agents

Local leaders, including community organizers, faith-based leaders, and neighborhood advocates, play a critical role in mobilizing support for homelessness initiatives. By empowering these leaders, organizations can foster a culture of support and shared responsibility at the grassroots level.

Strategies for Empowering Leaders:

Training and Resources: Provide leaders with the tools and resources they need to engage effectively, including workshops on advocacy and homelessness education.

Collaboration Opportunities: Offer platforms for local leaders to collaborate with service providers, policymakers, and other stakeholders.

Recognition and Support: Recognize the efforts of local leaders and provide continuous support to sustain their involvement.

Creating Community-Driven Solutions

Solutions that are created and driven by community members tend to be more sustainable and impactful. By empowering residents to co-design and implement homelessness initiatives, organizations can ensure that solutions reflect local priorities and are better accepted within the community.

Community-Driven Approaches:

Neighborhood Action Plans: Develop tailored action plans that address specific needs in different neighborhoods.

Peer Support Networks: Encourage the creation of peer-led support groups to provide mutual assistance.

Community Resource Hubs: Set up community centers or information hubs where residents can access resources and services.

Building Awareness and Reducing Stigma

Homelessness is often surrounded by stigma and misunderstanding. Educating communities about the causes of homelessness and the potential for recovery can foster compassion and reduce barriers to effective engagement.

Awareness Campaigns:

Public Education: Use local media, social media, and community events to share information and reduce stigma.

Humanizing Homelessness: Share personal stories and testimonials from those who have experienced homelessness to promote empathy.

School Outreach Programs: Engage youth through educational programs that teach empathy and social responsibility.

Engaging the Broader Community in Volunteering and Support

Volunteering offers community members a hands-on way to contribute to homelessness solutions. Engaging volunteers not only expands the reach of homelessness services but also helps volunteers understand the complexity of homelessness, fostering a culture of compassion.

Volunteer Programs:

Outreach Teams: Engage local volunteers in outreach efforts, such as distributing meals or providing companionship.

Skill-Based Volunteering: Encourage community members to volunteer their professional skills, such as counseling, healthcare, or legal assistance.

Fundraising Events: Organize community-driven events to raise funds and awareness for homelessness programs.

The Role of Faith-Based Organizations and Nonprofits

Faith-based organizations and nonprofits have a longstanding tradition of supporting homeless individuals through charitable work. Partnering with these groups can enhance community-based solutions, creating a wider safety net for individuals at risk of or experiencing homelessness.

Faith-Based and Nonprofit Partnerships:

Service Collaborations: Partner with these organizations to offer services like food distribution, emergency shelters, and counseling.

Community Outreach: Leverage these groups' established relationships within the community to reach vulnerable populations.

Advocacy Efforts: Work together to advocate for policy changes that benefit the homeless population.

Engaging Local Businesses and Private Sector Support

Businesses, both small and large, can contribute meaningfully to homelessness solutions through funding, resources, and job opportunities. Engaging local businesses fosters economic development and encourages the private sector to invest in community well-being.

Private Sector Initiatives:

Corporate Social Responsibility (CSR) Programs: Encourage businesses to allocate CSR funds toward homelessness programs.

Employment Opportunities: Collaborate with businesses to create job opportunities for individuals transitioning out of homelessness.

Donations and Sponsorships: Engage businesses in donating goods or sponsoring programs.

Measuring the Impact of Community Engagement

Evaluating the impact of community engagement efforts ensures that strategies are effective and helps identify areas for improvement. By measuring outcomes, organizations can adapt and refine their approaches to maximize the positive impact on homelessness.

Impact Measurement Approaches:

Qualitative Feedback: Gather feedback from community members to assess satisfaction and engagement levels.

Quantitative Metrics: Track metrics, such as participation rates, resource distribution, and volunteer hours.

Long-Term Outcomes: Measure the long-term impact of community engagement on homelessness rates and service accessibility.

Creating a Sustainable Culture of Community Involvement

To sustain the fight against homelessness, it is essential to create a culture of ongoing involvement where community members feel empowered to contribute. Building this culture requires regular engagement, transparency, and recognition of community contributions.

Strategies for Sustaining Involvement:

Regular Communication: Maintain open communication channels to keep the community informed and involved.

Celebrating Successes: Celebrate milestones and achievements with the community to reinforce commitment.

Continuous Engagement Opportunities: Offer ongoing opportunities for community members to participate in and support homelessness initiatives.

Empowering Communities to Drive Change

Empowering communities is vital for developing sustainable, localized solutions to homelessness in Los Angeles. By involving residents, leaders, businesses, and organizations, homelessness initiatives can benefit from a diverse array of insights, resources, and support. When communities feel a sense of ownership and responsibility, they are more likely to invest in lasting change.

Through community engagement, the fight against homelessness becomes a collective effort, bringing together individuals from all walks of life. By fostering empathy, reducing stigma, and empowering local leaders, communities can create a supportive environment where everyone is committed to making homelessness a thing of the past.

Chapter 23: Conclusion: Predicting Hope for a Brighter Future

In this final chapter, we bring together the insights, strategies, and possibilities explored throughout the book, emphasizing the role of data in not just predicting outcomes but in fostering hope and actionable change for those experiencing homelessness in Los Angeles. As homelessness remains one of the most complex social issues facing urban centers, the progress made through data-driven approaches reveals a path forward—one where informed interventions, community collaboration, and technological innovation work hand-in-hand to create a future with fewer people experiencing homelessness and more finding pathways to stability and dignity.

This chapter concludes the book and serves as a powerful call to action, reminding readers of the impact that data-driven approaches can have in addressing homelessness and inspiring them to contribute to this mission. By acknowledging the role of everyone—from policymakers to community members—in this journey, the conclusion reinforces the potential for a brighter future, grounded in both empathy and evidence.

Reflections on the Power of Data in Social Change

The journey through each chapter of this book illustrates how data, when used ethically and strategically, can transcend mere numbers to become a powerful tool for empathy, understanding, and transformation. Data enables policymakers, advocates, and community organizations to see beyond individual stories and spot trends, identify effective interventions, and allocate resources where they are most needed.

Key Takeaways:

Data as Insight: Transforming raw data into meaningful insights that can inform policies and drive results.

Data as Empathy: By revealing the systemic challenges faced by individuals, data allows us to approach solutions with compassion and urgency.

Data as Action: When turned into actionable strategies, data enables us to move from insight to tangible improvements in people's lives.

The Role of Innovation and Emerging Technologies

Emerging technologies, from predictive analytics to geospatial mapping, continue to open new possibilities for understanding and addressing homelessness. With advances in artificial intelligence, machine learning, and mobile data collection, cities like Los Angeles are poised to harness even more precise and effective tools for reducing homelessness.

Future Innovations:

AI and Machine Learning: Enhanced predictive models that can help prevent homelessness before it starts.

Mobile and Cloud-Based Platforms: Increasing accessibility to data in real-time, supporting rapid responses and emergency interventions.

Blockchain and Data Security: Securing personal information while enabling inter-agency collaboration and accountability.

A Vision for Lasting Change: Scaling and Replicating Success

The strategies and models developed in Los Angeles can serve as a blueprint for cities worldwide facing similar challenges. By sharing insights, best practices, and case studies, LA can become a model city for tackling homelessness through data-driven approaches.

Replicating Success:

Adapting LA's Model: Helping other cities implement data-based solutions customized to their unique needs.

Building a National Framework: Creating standards and protocols for data-driven homelessness initiatives across the U.S.

Global Partnerships: Engaging with cities around the world to exchange knowledge and support international efforts to address homelessness.

Strengthening Community Trust and Empowerment

A brighter future requires the sustained engagement and empowerment of local communities. By involving residents in data collection, analysis, and solutions, cities can create a more inclusive approach that recognizes the role of each individual in creating change.

Building a Culture of Inclusion:

Transparent Communication: Sharing data findings and decisions openly with the public to build trust.

Empowering Communities: Training community leaders and members in data literacy and advocacy.

Ongoing Engagement: Creating opportunities for communities to be active participants, not just recipients of solutions.

The Path Forward: Policy, Funding, and Long-Term Impact

Sustaining data-driven solutions requires supportive policies, consistent funding, and a commitment to measuring long-term outcomes. Policymakers, advocates, funders, and community members must work together to maintain the momentum and address emerging challenges.

Policy Recommendations:

Investment in Data Infrastructure: Ensuring that resources are allocated to maintain and improve data systems.

Flexible and Responsive Policies: Adapting policies as data reveals new insights or emerging needs.

Funding for Long-Term Solutions: Committing to sustained funding that goes beyond temporary relief to create lasting change.

Final Thoughts: Predicting Hope

Predicting hope is not about guaranteeing an end to homelessness overnight but about envisioning a future where data-driven actions reduce homelessness, support those in need, and foster a society where every person can thrive. By continuing to use data as both a guiding tool and a source of hope, we create a vision for a Los Angeles—and a world—where homelessness becomes a rare exception, not the rule.

Through collaboration, empathy, and innovation, the collective efforts of government agencies, nonprofits, private sectors, and communities have the potential to make a lasting impact. Together, we can move beyond merely addressing homelessness to preventing it and, ultimately, ending it.

Epilogue: A Call to Action

The journey to end homelessness in Los Angeles begins with each of us. Whether you are a policymaker, a data scientist, an advocate, or simply a concerned resident, your role is crucial. By supporting data-driven initiatives, advocating for fair policies, or volunteering time and resources, every effort counts.

As a Ghanaian author now residing in Los Angeles, the writer brings a unique cultural lens that values community cohesion, extended family support systems, and collective responsibility—elements deeply embedded in Ghanaian society. This perspective not only enhances the author's understanding of homelessness but also allows for a fresh viewpoint in addressing the complex issues around it in the U.S. His background offers potential value for American society as a whole, especially in a diverse city like Los Angeles, which faces a unique and challenging homelessness crisis. By integrating Ghanaian principles of community-centered support and culturally relevant approaches, the author can offer new solutions that resonate with both local and diverse populations.

In Ghana, homelessness is approached through a lens that emphasizes collective responsibility, where extended family and community networks often step in to prevent individuals from falling into destitution. This foundational belief is one the author believes could benefit American society in addressing homelessness. Unlike the prevalent individualistic approach in many Western societies, which tends to prioritize personal responsibility, Ghanaian values encourage shared responsibility and a communal safety net. By introducing these concepts, the author seeks to inspire American policymakers, social workers, and community leaders to consider more community-driven approaches, such as mentorship programs or "sponsor families," to support individuals at risk of homelessness.

Furthermore, Ghana's community savings groups and cooperative housing initiatives, which allow low-income families to collectively pool resources for housing or mutual support, offer a model that could resonate in the context of Los Angeles. The author believes that by adapting these financial models, American society could create more inclusive housing options for the homeless and low-income populations, perhaps even fostering a sense of community ownership and pride. This could prove particularly valuable in LA, where the housing crisis has deepened due to high costs and limited affordable housing options. Such initiatives could reduce housing costs and offer a sustainable support network, potentially alleviating some of the city's housing challenges.

The author's experience with culturally sensitive approaches to mental health in Ghana also brings a fresh perspective to LA's homelessness crisis. By valuing local traditions alongside mental health interventions, Ghanaian mental health strategies blend Western and indigenous practices. This approach, the author believes, could be highly beneficial in LA, especially in communities with diverse backgrounds where culturally relevant mental health support can foster stronger trust and engagement. Such tailored approaches can be instrumental in providing effective mental health services to homeless populations, particularly those who feel marginalized or disconnected from mainstream support systems.

Additionally, the author advocates for early intervention and preventive measures—a priority in Ghanaian social policy. This philosophy aligns with the concept of engaging at-risk individuals before they fall into homelessness, rather than waiting until they require intensive support. By focusing on community-level outreach and prevention, the author envisions creating sustainable pathways for reducing homelessness in LA. Initiatives aimed at families and youth could provide stability, offering a layer of protection and resilience to those on the edge of homelessness.

Finally, the author's Ghanaian heritage fosters a holistic view of data and policy, wherein policies are not only informed by quantitative data but also grounded in qualitative, community-driven insights. The author believes this dual approach can bring immense value to Los Angeles. While LA's current systems rely heavily on quantitative data, incorporating community-centered insights could offer a richer, more human understanding of homelessness. Data could then be used not just to identify risks but also to understand the community values, beliefs, and practices that shape people's lives and struggles.

The author's Ghanaian background therefore brings unique insights to American society, offering strategies that prioritize community resilience, cultural sensitivity, and preventive care. In Los Angeles, these approaches could prove especially beneficial, aligning with the city's diversity and complexity, and fostering a more inclusive, compassionate, and effective response to homelessness. Through these contributions, the author hopes to bridge Ghanaian values and American innovation, ultimately enhancing LA's capacity to address homelessness and supporting a broader dialogue on community and social responsibility across the U.S.

Ending homelessness is an ambitious goal that requires more than just government action; it demands a united community. It calls for a shared commitment to both compassion and evidence-based solutions. As individuals, we can champion the cause by pushing for accountability, investing in long-term initiatives, and ensuring the voices of those affected by homelessness are heard and respected.

Together, by blending innovation with compassion, we can predict a future where hope outshines despair, where every person has access to a safe home, and where homelessness is no longer an unsolvable crisis but a chapter that has been closed in our shared history.

Let this book be more than just a guide; let it be an invitation to join the journey toward lasting, positive change.

Glossary

1. Affordable Housing – Housing that is reasonably priced, allowing low-income households to meet basic living costs.

2. At-Risk Population – Groups more likely to face adverse outcomes, such as homelessness, due to factors like poverty or mental illness.

3. Behavioral Health – An umbrella term for mental health, substance abuse, and their effects on wellbeing.

4. Benchmarking – Setting performance standards based on best practices to evaluate effectiveness.

5. Big Data – Large and complex datasets that require advanced tools for processing and analysis.

6. Census Data – Demographic data collected by the government to provide population insights.

7. Chronic Homelessness – Long-term homelessness, often involving repeated episodes and linked to disabling conditions.

8. Community-Based Organization (CBO) – Nonprofits or other entities rooted in local communities working to address issues

like homelessness.

9. Continuum of Care (CoC) – A coordinated approach to provide homeless services from outreach to long-term housing.

10. Data-Driven Decision Making (DDDM) – Using data analysis to guide and inform policy and practice decisions.

11. Data Ethics – Principles governing the responsible and fair use of data, including privacy and consent.

12. Data Literacy – The ability to read, interpret, and use data effectively.

13. Demographic Analysis – Examination of population statistics, such as age and ethnicity, to understand social patterns.

14. Discharge Planning – Preparing individuals for stable housing and support upon release from institutions like hospitals or prisons.

15. Emergency Shelter – Temporary accommodations for people experiencing homelessness.

16. Evidence-Based Practices – Practices grounded in data and research showing proven outcomes.

17. Geospatial Analysis – Mapping data related to geographic locations, useful for identifying service gaps.

18. Homeless Management Information System (HMIS) – A data system used by homeless service providers to track client data and service needs.

19. Homelessness Prevention – Interventions designed to prevent individuals from becoming homeless.

20. Housing First – A model that prioritizes immediate housing for homeless individuals without preconditions.

21. Longitudinal Study – Research that follows subjects over a long period to observe changes and outcomes.

22. Machine Learning (ML) – A form of artificial intelligence enabling computers to learn patterns from data.

23. Marginalized Communities – Groups facing disadvantages due to systemic barriers, such as racial or economic inequities.

24. Mental Health Services – Programs aimed at providing therapy, medication, or other support for mental health needs.

25. Metrics – Quantitative measures used to assess program success and outcomes.

26. Outreach Programs – Efforts by service providers to connect with individuals on the streets or in shelters.

27. PIT Count (Point-in-Time Count) – An annual count of the homeless population on a single night, usually conducted in January.

28. Predictive Analytics – Using historical data to forecast future events or outcomes, like homelessness risk.

29. Public Policy – Laws and regulations enacted by governments to address societal issues.

30. Recidivism – The tendency of formerly homeless individuals to return to homelessness.

31. Rent Burden – When housing costs consume a high percentage of a household's income, increasing vulnerability to homelessness.

32. Resource Allocation – Distribution of resources, such as funding or services, to areas of greatest need.

33. Risk Assessment – Evaluation of factors that increase an individual's likelihood of experiencing homelessness.

34. Scatter-Site Housing – Dispersed housing units integrated into various neighborhoods, often part of Housing First programs.

35. Shelter Utilization Rate – The percentage of available shelter

beds that are occupied.

36. Social Determinants of Health – Economic and social conditions influencing health outcomes, including housing stability.

37. Sociodemographic Factors – Characteristics like age, gender, income, and education that impact an individual's life situation.

38. Stakeholder – Individuals or groups with an interest in addressing homelessness, including policymakers, nonprofits, and residents.

39. Substance Abuse Disorder – Addiction-related conditions impacting physical and mental health.

40. Supportive Housing – Housing paired with supportive services to help individuals maintain stability.

41. Sustainability – The ability to maintain or replicate programs long-term without exhausting resources.

42. Systemic Inequity – Institutional practices that result in unequal access to resources and opportunities for certain groups.

43. Transitional Housing – Temporary housing that helps individuals move from homelessness to permanent housing.

44. Trauma-Informed Care – An approach that recognizes the impact of trauma on individuals and adapts services to accommodate.

45. Underhoused – Individuals or families living in inadequate, overcrowded, or insecure housing.

46. Unsheltered Homelessness – People living on the streets or in places unfit for human habitation.

47. Urban Poverty – Poverty that occurs in urban settings, often associated with higher costs of living and fewer affordable housing options.

48. Veteran Homelessness – Homelessness affecting former military personnel, often with unique causes and needs.

49. Wraparound Services – Comprehensive services addressing multiple needs, including housing, mental health, and job training.

50. Zero Homelessness – A goal or state in which homelessness has been eliminated, with systems in place to prevent its recurrence.

Reference

Books

1. Burt, M. R. (2001). Helping America's homeless: Emergency shelter or affordable housing? The Urban Institute Press.

2. Culhane, D. P., & Metraux, S. (2008). A public health approach to ending homelessness. Journal of Public Health, 98(5), 781-787.

3. Elliott, M. (2014). The sociology of homelessness: Ethnographic studies of the homeless. Routledge.

4. Fitzpatrick, S., & Christian, J. (2006). The hidden truth about homelessness: Experiences of single homelessness in England. Crisis UK.

5. Gaetz, S., & Dej, E. (2017). A new direction: A framework for homelessness prevention. Canadian Observatory on Homelessness Press.

6. Goodman, L. A., & Thomson, D. M. (2021). The ethical use of predictive analytics in social work. NASW Press.

7. Grigsby, C., Baumann, D., Gregorich, S. E., & Roberts-Gray, C. (1990). Disaffiliation to entrenchment: A model for understanding homelessness. Journal of Social Issues, 46(4), 141-156.

8. Henry, M., Watt, R., Mahathey, A., Ouellette, J., & Sitler, A. (2019). The 2019 Annual Homeless Assessment Report (AHAR) to Congress. HUD.

9. Hopper, K., & Baumohl, J. (2005). Redefining the cursed word: A historical interpretation of American homelessness. Routledge.

10. Johnson, G., & Chamberlain, C. (2011). Are the homeless mentally ill? A case study of homelessness in Australia. Routledge.

11. Kuhn, R., & Culhane, D. P. (1998). Applying cluster analysis to test a typology of homelessness by pattern of shelter utilization: Results from the analysis of administrative data. American Journal of Community Psychology, 26(2), 207-232.

12. Link, B. G., & Phelan, J. (2001). Conceptualizing stigma. Annual Review of Sociology, 27, 363-385.

13. Mayock, P., & Bretherton, J. (2016). Women's homelessness in Europe. Palgrave Macmillan.

14. Metraux, S., & Culhane, D. P. (1999). Family dynamics, housing stability, and homelessness. The Urban Institute.

15. O'Flaherty, B. (2010). How to house the homeless. Russell Sage Foundation.

16. Rossi, P. H. (1989). Down and out in America: The origins of

homelessness. University of Chicago Press.

17. Shinn, M., & Khadduri, J. (2020). In the midst of plenty: Homelessness and what to do about it. Wiley.

18. Speer, J. (2018). Cities, police, and homelessness: Using theory to understand homelessness policy and street-level repression. Routledge.

19. Wright, T. (2016). Out of place: Homeless mobilizations, subcities, and contested landscapes. SUNY Press.

20. Wolch, J., & Dear, M. (2014). Landscapes of despair: From deinstitutionalization to homelessness. Princeton University Press.

Scholarly Articles

1. Baggett, T. P., O'Connell, J. J., Singer, D. E., & Rigotti, N. A. (2010). The unmet health care needs of homeless adults: A national study. American Journal of Public Health, 100(7), 1326-1333. https://doi.org/10.2105/AJPH.2009.180109

2. Barrow, S., Herman, D., Córdova, P., & Struening, E. (1999). Mortality among homeless shelter residents in New York City. American Journal of Public Health, 89(4), 529-534. https://doi.org/10.2105/AJPH.89.4.529

3. Bassuk, E. L., & Geller, S. (2006). The role of housing and services in ending family homelessness. Housing Policy Debate, 17(4), 781-806. https://doi.org/10.1080/10511482.2006.9521580

4. Byrne, T., Treglia, D., Culhane, D. P., Kuhn, J., & Kane, V. (2016). Predictors of homelessness among families and single adults after exit from homelessness prevention and rapid re-housing programs: Evidence from the Department of Veterans Affairs Supportive Services for Veteran Families Program. Housing Policy Debate, 26(1), 252-275. https://doi.org/10.1080/10511482.2015.1041257

5. Caton, C. L., Wilkins, C., & Anderson, J. (2007). People who experience long-term homelessness: Characteristics and interventions. Journal of Urban Health, 84(4), 524-534. https://doi.org/10.1007/s11524-007-9178-9

6. Culhane, D. P. (2008). The cost of homelessness: A perspective from the United States. European Journal of Homelessness, 2(1), 97-114.

7. Desmond, M., & Kimbro, R. T. (2015). Eviction's fallout: Housing, hardship, and health. Social Forces, 94(1), 295-324. https://doi.org/10.1093/sf/sov044

8. Elbogen, E. B., Sullivan, C., Wolfe, J., Wagner, H. R., & Beckham, J. C. (2013). Homelessness and money mismanagement in Iraq and Afghanistan veterans. American Journal of Public Health, 103(2), 248-254. https://doi.org/10.2105/AJPH.2013.301322

9. Fazel, S., Geddes, J. R., & Kushel, M. (2014). The health of homeless people in high-income countries: Descriptive epidemiology, health consequences, and clinical and policy recommendations. The Lancet, 384(9953), 1529-1540. https://doi.org/10.1016/S0140-6736(14)61132-6

10. Greenberg, G. A., & Rosenheck, R. A. (2008). Jail incarceration, homelessness, and mental health: A national study. Psychiatric Services, 59(2), 170-177. https://doi.org/10.1176/ps.2008.59.2.170

11. Henwood, B. F., & Katz, M. L. (2015). Housing stability, physical health status, and health care utilization among homeless adults in Los Angeles. Journal of Urban Health, 92(4), 785-795. https://doi.org/10.1007/s11524-015-9970-6

12. Hopper, E. K., Bassuk, E. L., & Olivet, J. (2010). Shelter from the storm: Trauma-informed care in homelessness services settings. The Open Health Services and Policy Journal, 3, 80-100. https://doi.org/10.2174/1874924001003020080

13. Johnsen, S., & Fitzpatrick, S. (2010). Revanchist sanitization or coercive care? The use of enforcement to combat street homelessness in the UK and USA. Urban Studies, 47(8), 1703-1723. https://doi.org/10.1177/0042098009356128

14. Jones, M. M. (2015). Does race matter in addressing homelessness? A review of the literature. World Medical & Health Policy, 7(1), 121-136. https://doi.org/10.1002/wmh3.132

15. Kuhn, R., & Culhane, D. P. (1998). Applying cluster analysis to test a typology of homelessness by pattern of shelter utilization: Results from the analysis of administrative data. American Journal of Community Psychology, 26(2), 207-232. https://doi.org/10.1023/A:1022176402357

16. Lee, B. A., Tyler, K. A., & Wright, J. D. (2010). The new homelessness revisited. Annual Review of Sociology, 36, 501-521. https://doi.org/10.1146/annurev-soc-070308-115940

17. Metraux, S., & Culhane, D. P. (2006). Homeless shelter use and reincarceration following prison release: Assessing the risk. Criminology & Public Policy, 5(2), 185–208. https://doi.org/10.1111/j.1745-9133.2006.00376.x

18. Padgett, D. K., Stanhope, V., Henwood, B. F., & Stefancic, A. (2011). Substance use outcomes among homeless clients with serious mental illness: Comparing Housing First with Treatment First programs. Community Mental Health Journal, 47(2), 227-232. https://doi.org/10.1007/s10597-009-9283-7

19. Pearson, C. L., Montgomery, A. E., & Locke, G. (2009). Housing stability among homeless individuals with serious mental illness participating in Housing First programs. Journal of Community Psychology, 37(3), 404-417. https://doi.org/10.1002/jcop.20301

20. Petrovich, J., & Robbins, D. L. (2017). Predictors of homelessness among veterans: A systematic review. Social Work in Public Health, 32(6), 412-429. https://doi.org/10.1080/19371918.2017.1316676

21. Pleace, N. (2016). Exclusion by definition: The underrepresentation of women in European homelessness statistics. European Journal of Homelessness, 10(1), 163-183.

22. Raoult, D., & Arzounian, D. (2012). Health care needs of the homeless population. International Journal of Environmental Research and Public Health, 9(4), 941-956. https://doi.org/10.3390/ijerph9040941

23. Rog, D. J., & Buckner, J. C. (2007). Homeless families and children. In D. Dennis, G. Locke, & J. Khadduri (Eds.), Toward understanding homelessness: The 2007 national symposium on homelessness research (pp. 1-32). U.S. Department of Housing and Urban Development.

24. Rosenheck, R., & Fontana, A. (1994). A model of homelessness among male veterans of the Vietnam War generation. American Journal of Psychiatry, 151(3), 421-427. https://doi.org/10.1176/ajp.151.3.421

25. Shlay, A. B., & Rossi, P. H. (1992). Social science research and contemporary studies of homelessness. Annual Review of Sociology, 18, 129-160. https://doi.org/10.1146/annurev.so.18.080192.001021

26. Stergiopoulos, V., Gozdzik, A., Nisenbaum, R., & Durbin, J. (2015). Effectiveness of Housing First programs in retaining homeless people with mental illness. Health Affairs, 34(7), 1166-1173. https://doi.org/10.1377/hlthaff.2014.1528

27. Tsai, J., & Rosenheck, R. A. (2015). Risk factors for homelessness among US veterans. Epidemiologic Reviews, 37(1), 177-195. https://doi.org/10.1093/epirev/mxu004

28. Tsemberis, S. (2010). Housing First: Ending homelessness,

promoting recovery, and reducing costs. In I. Gould Ellen & B. O'Flaherty (Eds.), How to house the homeless (pp. 37-56). Russell Sage Foundation.

29. Turner, N., & Moran, L. (2020). Using predictive analytics to address homelessness. Journal of Social Service Research, 46(3), 325-338. https://doi.org/10.1080/01488376.2020.1717148

30. Williams, A. M., & Hall, C. M. (2014). Homelessness, health, and policy: A European perspective. European Journal of Public Health, 24(2), 169-176. https://doi.org/10.1093/eurpub/ckt101

CHAPTER 18: SCALING IMPACT: REPLICATING SUCCESS AND ADDRESSING CHALLENGES

As we work toward solving the homelessness crisis in Los Angeles, it is essential to not only create effective, data-driven solutions but also to scale these efforts across the city and beyond.

Replicating success and addressing the challenges that arise when expanding programs is key to ensuring that the progress made can be sustained and that more individuals experiencing homelessness can benefit from these solutions.

Scaling impact requires both strategic planning and the ability to overcome challenges that are inherent to expanding efforts that involve diverse communities, stakeholders, and funding sources. In this chapter, we will explore the strategies for replicating successful homelessness interventions, the challenges encountered during scaling, and the methods for overcoming these obstacles.

Replicating Success: Key Strategies

To scale the impact of homelessness interventions, it is important to first identify the elements that contributed to the success of programs. These key strategies allow for the replication of effective solutions in different regions or neighborhoods and across larger populations.

Standardizing Best Practices

Replicating successful models requires establishing a set of best practices that can be implemented consistently across different communities or jurisdictions. Whether it is the Housing First approach, Rapid Rehousing, or outreach programs, best practices help guide new initiatives in the right direction.

A well-documented implementation plan ensures that the core principles that contributed to the success of the original program are maintained when the program is expanded. This plan should cover operational procedures, data collection protocols, staff training, community engagement strategies, and service delivery methods.

Leveraging Data for Informed Replication

Data is crucial for identifying which interventions work best in certain contexts. By analyzing the success factors that led to positive outcomes, including housing retention rates, health improvements, and employment gains, we can determine the elements of the program that are most effective.

Replicating success also requires understanding the demographics and local contexts of other areas where the program will be expanded. Data-driven insights into the social, economic, and environmental conditions of different neighborhoods will allow for a tailored approach to replicating successful models.

Fostering Cross-Sector Partnerships

A collaborative approach is essential for scaling homelessness solutions. The success of many homelessness programs is the result of partnerships among government agencies, nonprofit organizations, businesses, and local communities. Replicating these efforts in other regions requires the same level of coordination and buy-in from all sectors involved.

Key partnerships can be forged between public health agencies, housing authorities, mental health services, and employment agencies. By working together, these sectors can pool resources, share data, and provide comprehensive services to individuals experiencing homelessness.

Ensuring Financial Sustainability

Scaling homelessness programs also involves securing long-term funding. Whether it's federal, state, or local government support, philanthropic donations, or revenue from service fees, a sustainable funding model is essential for the success of expansion efforts.

Replicating success may require advocacy for increased funding or the establishment of innovative financial models, such as public-private partnerships. These partnerships can help reduce the reliance on government funding and involve stakeholders who are invested in the long-term success of the program.

Training and Capacity Building

As homelessness programs scale, it is essential to train staff and increase the capacity of service providers. Training programs for case managers, outreach workers, and other service providers will ensure that the quality of service is maintained as the program expands. Additionally, building capacity within local communities to manage and support these initiatives can help sustain the program's success.

Challenges in Scaling Homelessness Programs

While scaling successful interventions is necessary, there are numerous challenges associated with expansion. Understanding these challenges helps to develop strategies for addressing them and ensuring that scaling efforts are effective and sustainable.

Resource Constraints

As programs expand, it is inevitable that the demand for resources—whether financial, human, or infrastructural—will increase. Ensuring adequate funding, staff, and physical space for service delivery can be challenging, especially in cities with large homeless populations and limited budgets.

Scaling often involves not just increasing resources but also ensuring that existing resources are used effectively. Efficiently allocating resources, such as staff time, financial investments, and housing units, can help mitigate resource constraints.

Community Resistance and NIMBYism

When homelessness programs are replicated in new neighborhoods or regions, they may face opposition from local communities. Not In My Backyard (NIMBY) sentiments often arise, as some individuals and groups resist efforts to house homeless populations in their area. This resistance can be rooted in misconceptions about homelessness, fear of crime, or concerns over property values.

Addressing community resistance requires clear communication and engagement. Community members must be educated about the benefits of homelessness programs, including the positive social and economic impacts of reducing homelessness. Building trust and fostering open dialogue can help overcome NIMBYism.

Maintaining Service Quality

As programs expand, there is a risk that the quality of services may diminish if not carefully managed. Ensuring consistency in the level of care and support provided to individuals experiencing homelessness is essential for the long-term success of scaling efforts. It is important to monitor the quality of services regularly and make necessary adjustments to prevent any degradation of service delivery.

Maintaining quality may require the introduction of standardized protocols across service providers to ensure that all individuals receive the same level of care and support, regardless of where they live or which program they participate in.

Political and Policy Barriers

Homelessness is often a highly politicized issue, and changes in political leadership or policy priorities can disrupt efforts to scale programs. Shifting priorities at the local, state, or federal levels can lead to funding cuts, changes in regulations, or the de-prioritization of homelessness interventions.

To address these challenges, it is essential to build strong, bipartisan support for homelessness programs. Advocacy and policy engagement efforts should focus on demonstrating the success and effectiveness of data-driven homelessness interventions in reducing the problem, thereby ensuring continued political support.

Coordination Across Different Agencies

Scaling homelessness programs often involves collaboration across many different organizations, which can be challenging. Different agencies may have varying priorities, resources, and operational processes, making it difficult to coordinate efforts effectively.

To overcome this challenge, it is important to create memorandums of understanding (MOUs) or formal agreements that establish roles, responsibilities, and expectations for each participating agency. Regular communication and joint planning sessions are essential to keep all stakeholders aligned as the program scales.

Strategies for Overcoming Scaling Challenges

To effectively scale homelessness programs and overcome the challenges identified above, there are several key strategies that can be employed.

Adaptability and Flexibility

Scaling programs requires adaptability. Different neighborhoods, regions, and populations may have unique needs, so it is important to tailor interventions and services to the specific context. Programs must be flexible enough to adapt to different geographic, economic, and social conditions while still maintaining the core principles of the original intervention.

Engagement and Advocacy

Engaging with community stakeholders early and often can help overcome resistance and ensure that local communities understand the value of homelessness programs. This engagement should also include advocacy efforts to secure financial support and policy changes that enable program scaling.

Continuous Evaluation and Improvement

Scaling should not be seen as a one-time event but as a continuous process. Regular evaluation through performance metrics and feedback loops allows for ongoing improvement. By collecting data on the effectiveness of the scaled programs and making adjustments as needed, service providers can ensure that their impact continues to grow.

Building Local Leadership and Ownership

Empowering local leaders and stakeholders to take ownership of the program's expansion can increase the likelihood of success. Local champions who understand the community's unique needs and dynamics can drive the program's expansion and ensure its sustainability.

Scaling impact requires careful planning, strong partnerships, and the ability to address the challenges that come with expansion. By replicating successful homelessness interventions, leveraging data, and fostering collaboration, we can ensure that the progress made in reducing homelessness can be extended to more individuals and communities. However, scaling is not without its obstacles, and it is important to approach it strategically, continuously adapting and evaluating the programs as they grow. Through these efforts, we can make meaningful strides toward solving the homelessness crisis not just in Los Angeles, but in other regions as well, creating a more equitable and compassionate society for all.

CHAPTER 19:
EMERGING TRENDS
AND TECHNOLOGIES:
OPPORTUNITIES AND
CHALLENGES

*Emerging trends and technologies
are playing an increasingly vital
role in shaping the landscape
of homelessness prevention,
intervention, and service delivery.*

These trends not only offer new opportunities for improving the efficiency and effectiveness of programs but also present a set of challenges that must be navigated carefully.

In this chapter, we will explore the latest technological advancements and trends influencing the fight against homelessness, the potential they offer for solving this complex issue, and the challenges they introduce. From advancements in artificial intelligence (AI) and data analytics to the application of blockchain and digital platforms, these technologies have the potential to reshape homelessness interventions, but they must be harnessed thoughtfully to ensure their benefits are realized for all populations.

Emerging Trends and Technologies Shaping Homelessness Solutions

Artificial Intelligence and Machine Learning

Artificial intelligence (AI) and machine learning (ML) are increasingly being used to analyze large sets of data, predict trends, and identify the most effective interventions for individuals experiencing homelessness. These technologies offer powerful tools for creating predictive models that can help communities anticipate homelessness risk and intervene proactively.

For instance, machine learning algorithms can process historical data on individuals' pathways into homelessness and predict who is at risk, enabling service providers to offer targeted interventions before a person loses their home. Additionally, AI-powered chatbots and virtual assistants can assist case managers by providing real-time information and support for clients.

Opportunity: Improved prediction of homelessness risk, tailored service delivery, and automated support for case management.

Challenge: Ensuring that AI models are accurate and free from biases, especially in communities with complex socio-economic dynamics.

Blockchain for Data Security and Transparency

Blockchain technology has the potential to revolutionize how data is managed and shared across different agencies and organizations working to address homelessness. With its decentralized nature, blockchain can ensure that personal data is securely stored and shared, providing greater privacy and control to individuals experiencing homelessness while enabling seamless coordination among service providers.

Blockchain can also be used for the management of resources, such as housing vouchers, social benefits, or financial aid, ensuring transparency, reducing fraud, and providing a clear, auditable record of transactions.

Opportunity: Increased data security, transparency in resource distribution, and streamlined coordination between agencies.

Challenge: Implementing blockchain in the homelessness sector requires overcoming technical barriers, including integration with existing systems and ensuring accessibility for all users.

Mobile and Digital Platforms for Service Access

Mobile apps and digital platforms are transforming how individuals access services and support. For those experiencing homelessness, access to services can often be challenging due to mobility issues, lack of stable internet, or limited physical infrastructure. Digital platforms can bridge this gap by offering mobile-based solutions that allow individuals to connect with resources such as shelters, food programs, healthcare services, and job opportunities.

Apps can also provide real-time information, such as available shelter space or transportation options, and enable direct communication with case managers or service providers.

Opportunity: Enhanced access to services and information, improved communication between service providers and clients, and increased self-sufficiency for individuals experiencing homelessness.

Challenge: Digital access may be limited among certain homeless populations, and there may be challenges in ensuring the security and privacy of sensitive personal information.

Data Analytics and Real-Time Monitoring

Data analytics is playing an essential role in monitoring homelessness trends and optimizing service delivery. By utilizing real-time data, service providers can track the availability of resources, monitor the effectiveness of interventions, and quickly adjust strategies based on emerging patterns. Real-time monitoring allows for better coordination, ensuring that services are available when and where they are needed most.

For instance, data dashboards can track shelter occupancy rates, medical services utilization, and the status of housing placements, helping service providers allocate resources dynamically and prevent bottlenecks in service delivery.

Opportunity: Enhanced decision-making, real-time resource allocation, and better understanding of homelessness trends.

Challenge: The need for accurate, up-to-date data and the challenge of integrating data from multiple sources without overwhelming caseworkers or service providers.

Virtual and Augmented Reality for Housing Solutions

Virtual reality (VR) and augmented reality (AR) technologies are beginning to be applied in innovative ways to address housing issues. VR can be used to offer virtual tours of housing options, enabling individuals experiencing homelessness to view potential homes remotely. AR applications can assist in visualizing potential modifications to housing spaces, such as for accessibility or safety purposes.

Moreover, VR can serve as a training tool for both service providers and individuals experiencing homelessness, helping case managers practice various scenarios and improve their skills in interacting with clients.

Opportunity: Enhanced access to housing options, improved training for service providers, and the ability to remotely view housing opportunities.

Challenge: The cost and availability of VR/AR technology, as well as the need to ensure that these technologies are user-friendly and accessible to individuals who may have limited technological literacy.

Smart Cities and Internet of Things (IoT)

The concept of smart cities is being explored as a way to create more connected, sustainable, and efficient urban environments. Smart city technologies, powered by the Internet of Things (IoT), can be used to improve homelessness interventions. For instance, IoT sensors can monitor shelter conditions, track air quality, and optimize the delivery of resources like food or medical supplies.

Additionally, smart infrastructure, such as public Wi-Fi and smart transportation systems, can provide greater access to essential services for individuals experiencing homelessness, helping them navigate the challenges of urban living.

Opportunity: Greater integration of technology in urban spaces, improved resource delivery, and enhanced living conditions for individuals experiencing homelessness.

Challenge: Ensuring that the benefits of smart city technologies are equitably distributed and accessible to all members of the community, including those experiencing homelessness.

Challenges in Leveraging Emerging Technologies

While the opportunities presented by emerging technologies are vast, there are several challenges that need to be addressed to ensure that these innovations effectively serve homeless populations. These challenges include:

Equity and Access

Not all individuals experiencing homelessness have access to smartphones, computers, or reliable internet. Digital divides, especially in low-income and homeless communities, can limit the effectiveness of technology-driven solutions. Ensuring that technological interventions are accessible to all individuals—regardless of their access to devices or internet connectivity—is essential to the success of these innovations.

Data Privacy and Security

As more personal data is collected through digital platforms, mobile apps, and AI-powered tools, safeguarding the privacy and security of this data becomes critical. Homeless individuals are often vulnerable to identity theft, fraud, or misuse of their personal information. Implementing strong data protection measures is vital to maintaining trust and ensuring that individuals' sensitive information is kept secure.

Cost and Implementation Barriers

The implementation of emerging technologies can be costly, both in terms of initial investments and ongoing maintenance. For cash-strapped service providers and government agencies, funding and resources may not be sufficient to incorporate high-tech solutions at scale. Moreover, technological solutions require training and support for both service providers and homeless individuals, which can add to the cost and complexity of implementation.

Integration with Existing Systems

Integrating new technologies into existing systems and workflows can be complex. Homelessness service providers often rely on outdated systems or disconnected databases, making it difficult to introduce and maintain new technological tools. Overcoming technical barriers to integration is a key challenge when scaling technological solutions in the homelessness sector.

Ethical Considerations

The use of AI, blockchain, and other technologies raises important ethical considerations. For example, algorithmic bias in AI models could exacerbate inequalities or unfairly target certain populations. Similarly, blockchain applications need to ensure that the use of personal data does not inadvertently stigmatize or harm individuals experiencing homelessness.

Opportunities for Innovation and Future Trends

As technology continues to evolve, there are several emerging trends that hold promise for addressing homelessness in new and innovative ways. These trends include:

AI-Powered Predictive Tools for early intervention and personalized services.

5G Networks and Connectivity to ensure reliable access to digital platforms and resources for homeless individuals.

Smart Housing Solutions such as modular, affordable housing built with technology-driven efficiency.

Wearable Technologies that track health and safety, alerting service providers in real-time to potential issues.

As we move forward, it is important to remain open to new ideas and solutions. By embracing these emerging technologies thoughtfully and ethically, we can create a future where homelessness is not only addressed more effectively but ultimately eradicated.

Emerging trends and technologies offer exciting opportunities for improving how we address homelessness, making interventions more targeted, efficient, and accessible. However, with these opportunities come significant challenges that must be navigated carefully. By addressing equity, privacy, and integration concerns, and ensuring that technological solutions are used ethically and sustainably, we can harness the power of innovation to drive meaningful change in the fight against homelessness. As we continue to explore the potential of these technologies, we must keep in mind that the ultimate goal is to create a system that works for everyone, leaving no one behind.

CHAPTER 20: POLICY AND ADVOCACY: USING DATA TO INFORM DECISION-MAKING

Data a powerful tool for shaping policy,driving advocacy and influencing decesion making.

In the effort to address homelessness, data is not just a tool for service delivery but also a powerful instrument for shaping policy, driving advocacy, and influencing decision-making at local, state, and national levels. By leveraging data effectively, we can identify gaps in existing systems, demonstrate the impact of homelessness interventions, and advocate for the resources and reforms necessary to create lasting solutions. This chapter explores the role of data in informing policy and advocacy, providing insights on how data can drive systemic change and improve outcomes for individuals experiencing homelessness.

The Role of Data in Shaping Homelessness Policy

Homelessness is a complex, multifaceted issue that requires comprehensive and coordinated responses across various sectors, including housing, healthcare, social services, and employment. Data serves as the foundation for designing, implementing, and evaluating policies aimed at reducing homelessness. Policymakers need accurate, timely, and reliable data to understand the scope of the problem, identify trends, and assess the effectiveness of existing programs.

Informing Policy Development

Data can inform the development of policies that address the root causes of homelessness, including lack of affordable housing, economic inequality, mental health issues, and substance abuse. By analyzing data from various sources, policymakers can craft targeted policies that respond to the unique needs of different homeless populations, such as families, veterans, individuals with disabilities, or those experiencing chronic homelessness.

For example, data on the availability of affordable housing can be used to advocate for zoning changes or increased funding for housing subsidies. Data on the prevalence of mental health or substance use disorders among the homeless population can lead to the creation of more comprehensive healthcare services tailored to these individuals' needs.

Opportunity: Data enables policymakers to design evidence-based policies that are more likely to achieve positive outcomes and effectively address homelessness.

Challenge: Data must be comprehensive and accurately reflect the diverse factors contributing to homelessness, or policies may fail to address underlying causes.

Assessing Program Effectiveness

Data can also be used to evaluate the effectiveness of homelessness programs and services. By tracking key performance indicators (KPIs), such as shelter occupancy rates, housing retention, employment outcomes, and health improvements, policymakers can assess which programs are working and which need to be adjusted or discontinued.

For example, if a rapid rehousing program shows high success rates in quickly placing individuals into stable housing, data can be used to advocate for the expansion of that program. Conversely, if a program is found to be ineffective, data can highlight areas for improvement or inform the decision to reallocate resources to more successful initiatives.

Opportunity: Data-driven evaluations allow for continuous improvement of homelessness programs, ensuring that resources are used effectively to achieve the best possible outcomes.

Challenge: Data collection and analysis must be consistent and standardized across programs to ensure comparability and accurate assessments.

Advocating for Policy Reform

Advocacy is a critical part of driving systemic change, and data plays a central role in this process. By presenting compelling data that highlights the severity of homelessness and the effectiveness of certain interventions, advocates can persuade policymakers, funders, and the public to prioritize homelessness as a key issue.

Data can be used to demonstrate the economic costs of homelessness, such as the strain on emergency services, healthcare, and criminal justice systems, as well as the potential savings from prevention and intervention programs. For instance, studies showing that every dollar spent on homelessness prevention saves multiple dollars in healthcare and criminal justice costs can be a powerful tool for advocating for increased funding for prevention and rapid rehousing programs.

Opportunity: Data can provide the evidence needed to build a strong case for policy reforms and secure funding for homelessness programs.

Challenge: Advocacy efforts must ensure that data is presented in an accessible and persuasive manner to reach diverse audiences, including policymakers, donors, and the public.

Types of Data Essential for Policy and Advocacy

Different types of data are necessary for informing effective policies and advocacy efforts. Below are some key types of data that play a significant role in addressing homelessness:

Demographic Data

Understanding the demographics of the homeless population is essential for designing targeted interventions. Data on factors such as age, gender, race, family composition, and immigration status can help policymakers develop policies that address the specific needs of different subpopulations.

Example: A growing number of homeless youth or families with children may prompt the development of more family-centered housing programs and child welfare services.

Service Utilization Data

Data on the use of homelessness services—such as shelters, food programs, healthcare, and employment assistance—helps policymakers identify gaps in service delivery and allocate resources more efficiently. Tracking how frequently services are accessed and by whom can reveal patterns and guide decisions on where to expand or improve services.

Example: If a region experiences high demand for shelters but a low success rate in housing placements, policymakers may prioritize expanding rapid rehousing initiatives or increasing the availability of long-term supportive housing.

Housing Data

Housing data, including the availability of affordable housing, vacancy rates, rent levels, and the number of housing subsidies or vouchers issued, is critical for informing policies aimed at increasing access to stable housing for homeless individuals.

Example: Data on the shortage of affordable housing can be used to advocate for new affordable housing projects, zoning reforms, or the expansion of housing voucher programs.

Health and Social Services Data

Homelessness is often intertwined with health and social issues, such as mental health conditions, substance use disorders, chronic illness, and trauma. Data on health and social services utilization can help identify trends and gaps in service provision, enabling policymakers to create integrated health and housing solutions.

Example: Data on the prevalence of mental health disorders among homeless populations can support the creation of specialized housing-first programs that provide both stable housing and mental health services.

Economic and Employment Data

Employment and income data is key for understanding the economic factors contributing to homelessness. Data on unemployment rates, wages, and income inequality can inform policies aimed at addressing the economic drivers of homelessness.

Example: If data shows a high correlation between unemployment and homelessness in certain regions, policies may focus on job training programs, wage increases, and employment opportunities for homeless individuals.

Using Data to Influence Policy at Various Levels

Data-driven advocacy is essential for influencing policy decisions at the local, state, and national levels. Here's how data can be used to affect policy at each level:

Local Level

At the local level, data is often used to advocate for changes in zoning laws, the development of affordable housing, or the allocation of funding for homelessness programs. Local government officials rely on data to identify the most pressing needs in their communities and prioritize interventions accordingly.

Example: In cities with high rates of homelessness, data showing the effectiveness of permanent supportive housing programs can be used to encourage local governments to prioritize funding for these initiatives.

State Level

State governments often control funding for homelessness programs and have the ability to enact policies that affect the broader housing and social welfare systems. Data on statewide homelessness trends, such as regional disparities in housing availability or the effectiveness of eviction prevention programs, can be used to advocate for state-level reforms.

Example: Data showing the success of a particular state's eviction prevention program can be used to advocate for the replication of that model in other regions of the state.

National Level

At the national level, data is used to inform large-scale policy decisions, including federal funding allocations, the creation of national initiatives to combat homelessness, and the development of comprehensive homelessness prevention strategies.

Example: Data on the high costs of chronic homelessness, including emergency healthcare and justice system involvement, can be used to argue for greater federal investment in homelessness prevention and housing-first programs.

Challenges in Using Data for Policy and Advocacy

While data is an invaluable tool for shaping policy and advocacy efforts, several challenges can hinder its effectiveness:

Data Gaps and Quality Issues

Incomplete or poor-quality data can undermine advocacy efforts. Inaccurate or missing data on homelessness trends, service utilization, or program effectiveness can lead to misguided policy decisions.

Privacy and Confidentiality Concerns

Data collection and use must balance the need for comprehensive information with the protection of individuals' privacy. Ensuring that data is collected and shared responsibly is essential to maintaining public trust and safeguarding vulnerable populations.

Political and Bureaucratic Barriers

In some cases, political or bureaucratic obstacles can delay or prevent the use of data to inform policy decisions. Decision-makers may be resistant to adopting data-driven solutions if they conflict with existing policies or political priorities.

Best Practices for Data-Driven Advocacy and Policy Implementation

To maximize the impact of data in advocacy and policymaking, organizations and advocates must adhere to best practices that enhance the reliability, transparency, and accessibility of data. The following strategies ensure that data serves its full potential in the fight against homelessness:

1: Data Standardization and Consistency

One of the critical aspects of successful data-driven policymaking is the standardization of data. This ensures that data collected from different sources, regions, or organizations can be compared and analyzed effectively. Standardized data allows policymakers to make informed decisions based on reliable, apples-to-apples comparisons.

Actionable Strategy: Advocate for the adoption of common data standards across different agencies, service providers, and jurisdictions involved in homelessness services. This could include using standardized case management tools, consistent definitions for homelessness, and uniform reporting formats.

Impact: Consistent data helps ensure that interventions are aligned with regional and national standards, facilitating collaboration and scaling up successful approaches.

2: Data Transparency and Accessibility

For data to be impactful in policy and advocacy, it must be easily accessible to those who need it. This means not only making data available to policymakers but also to community organizations, the public, and people directly affected by homelessness.

Actionable Strategy: Develop online platforms and dashboards where homelessness data can be shared in a user-friendly format. These tools can help communities, researchers, and policymakers access real-time information and analyze trends in homelessness.

Impact: When data is accessible and transparent, it fosters accountability, allows for public scrutiny, and encourages informed decision-making at all levels.

3: Collaboration Across Sectors

The complex nature of homelessness means that addressing it requires collaboration across multiple sectors, including housing, healthcare, education, and employment. Effective data-driven policy solutions must be the product of cross-sector collaboration.

Actionable Strategy: Create formal partnerships between

government agencies, nonprofit organizations, businesses, healthcare providers, and academic institutions. These partnerships can help pool data from diverse sources, ensuring a comprehensive understanding of homelessness.

Impact: Collaborating across sectors allows for holistic interventions that address the multiple facets of homelessness, ensuring that solutions are comprehensive and sustainable.

4: Engaging Affected Communities in Data Collection

While it is crucial to use data to inform decision-making, it is equally important to involve the homeless population in the data collection process. Engaging individuals with lived experience ensures that the data collected is reflective of their needs and experiences, rather than being driven solely by external perspectives.

Actionable Strategy: Incorporate feedback from individuals experiencing homelessness in the design and implementation of data collection efforts. Use participatory research methods, such as interviews, focus groups, and community surveys, to gather insights directly from those affected.

Impact: This approach ensures that the data is more accurate and relevant to the lived experiences of homeless individuals, which in turn can help create policies that are more empathetic and effective.

5: Real-Time Data for Rapid Response

In order to be most effective, data must not only be accurate but also timely. Having real-time data allows for quick responses to emerging homelessness trends or immediate needs.

Actionable Strategy: Implement systems that allow for the real-time collection and analysis of data, such as mobile applications for service providers, GPS tracking for outreach teams, or automated data-entry systems.

Impact: Real-time data allows for quicker adjustments to service delivery, more immediate interventions, and better crisis management.

Data is a powerful tool in the fight against homelessness. It provides the evidence necessary to inform policymaking, evaluate the effectiveness of interventions, and advocate for the resources needed to address the issue. By leveraging data effectively

Harnessing Big Data and Technology in Homelessness Policy

As technology continues to evolve, new opportunities to leverage big data and advanced analytics for homelessness policy emerge. From predictive modeling to geospatial analysis, the possibilities for using data to tackle homelessness are expanding rapidly. Below are some cutting-edge technologies that can be used to influence policy and advocacy efforts:

Predictive Analytics and Forecasting

Predictive analytics uses historical data and advanced algorithms to forecast future trends and behaviors. By analyzing factors such as housing market trends, economic conditions, and previous service usage, predictive models can identify individuals or populations at risk of homelessness before they become homeless.

Actionable Strategy: Use predictive models to identify high-risk individuals and offer proactive interventions, such as eviction prevention programs, early case management, or financial assistance.

Impact: Predictive analytics can help prevent homelessness before it occurs, enabling policymakers and service providers to address issues early and avoid costly interventions later.

Geospatial Data and Mapping

Geospatial data analysis involves using geographic information systems (GIS) to visualize and analyze spatial data. This can be particularly useful for identifying patterns in homelessness and service access, helping policymakers determine the best locations for resource allocation.

Actionable Strategy: Utilize GIS tools to map areas with high concentrations of homelessness, identify gaps in services, and prioritize areas for new housing or shelter development.

Impact: GIS mapping allows for smarter resource allocation, ensuring that interventions are implemented in areas where they are most needed.

Artificial Intelligence and Machine Learning

Artificial intelligence (AI) and machine learning (ML) can be used to automate data processing, identify patterns, and predict outcomes. These technologies can improve the speed and accuracy of data analysis, helping policymakers make data-driven decisions more quickly.

Actionable Strategy: Implement AI-powered data analysis tools that can sift through large datasets, identify trends, and predict the success of different homelessness interventions.

Impact: By automating routine data analysis, AI and ML can help policymakers and advocates make faster, more informed decisions about how to address homelessness.

Challenges and Ethical Considerations in Data-Driven Policy

Despite the immense potential of data to inform policy and advocacy, there are several challenges and ethical considerations that must be addressed to ensure that data is used responsibly:

Privacy and Confidentiality

Homeless individuals are often vulnerable to exploitation and discrimination. Ensuring the privacy and confidentiality of the data collected is critical to building trust with the community and protecting individuals' rights.

Solution: Implement strict data protection protocols, such as encryption, de-identification, and secure data storage, to ensure that sensitive information is protected.

Bias and Data Quality

Data can sometimes reflect biases in its collection or interpretation, which can lead to skewed outcomes. It is essential to ensure that data is representative and unbiased, especially when it comes to marginalized populations such as homeless individuals.

Solution: Engage with communities affected by homelessness to ensure that their perspectives and experiences are accurately represented in data collection efforts. Regularly audit data for biases and gaps.

Accessibility and Equity

There is a risk that data-driven policies may disproportionately benefit those who already have access to resources, leaving out marginalized communities. Ensuring that data policies are equitable and inclusive is essential for addressing the root causes of homelessness.

Solution: Advocate for policies that ensure data collection and analysis efforts are inclusive, reflecting the diversity of experiences within the homeless population.

Data as a Catalyst for Change

Data has the power to transform how we address homelessness, from shaping policies and advocacy efforts to optimizing service delivery and identifying new opportunities for intervention. By employing data strategically and ethically, we can create more effective, sustainable solutions to homelessness, improving outcomes for individuals and communities.

To achieve this, we must prioritize data collection, standardization, and accessibility, invest in new technologies, and foster collaboration across sectors. With these efforts, data can be harnessed to create systemic change, drive policy reforms, and ultimately end homelessness. The future of homelessness prevention and intervention lies in data-driven strategies that are responsive, adaptive, and inclusive, ensuring that everyone has a pathway to stability and self-sufficiency

CHAPTER 21:
SUSTAINABILITY AND FUNDING: ENSURING LONG-TERM IMPACT

In the journey to end homelessness, achieving sustainable impact is crucial.

While immediate relief efforts offer essential short-term support, creating long-term change requires a commitment to sustainable strategies and funding models. This chapter will explore how data-driven solutions provide a foundation for lasting impact, how innovative funding models can support these efforts, and why a collaborative approach is essential for success.

The Role of Data in Securing Sustainability

Data transforms the approach to homelessness by making the invisible visible. Through data, we can see which populations are most vulnerable, what services they need, and where resources are most effectively allocated. Importantly, data helps track impact over time, enabling us to refine strategies as needs evolve. A data-driven approach doesn't just react to crises but anticipates them, helping allocate resources proactively.

For instance, predictive analytics can identify individuals at risk of homelessness, allowing service providers to offer targeted assistance before they lose stable housing. By gathering and analyzing data on the effectiveness of various interventions —such as job training, mental health services, and affordable housing initiatives—organizations can focus resources on solutions proven to yield lasting results. When funding decisions are guided by data, every dollar spent contributes toward sustainable, impactful change.

Funding Models for Homelessness Prevention

Achieving sustainability requires more than one funding source; it demands diverse funding models to secure a steady flow of resources. Several funding strategies have proven effective for homelessness initiatives:

1. Government Grants

Federal, state, and local grants provide critical support for programs addressing homelessness. Many of these grants focus on housing, mental health services, job training, and other essential services. Applying for these grants requires detailed data on program effectiveness, making data-driven insights invaluable.

2. Nonprofit Partnerships

Partnering with established nonprofits allows organizations to access additional resources and leverage the nonprofit's expertise and reputation. These partnerships often attract further support from donors who trust the nonprofit's commitment and accountability.

3. Philanthropy and Donations

Philanthropic organizations and individual donors play a significant role in funding social impact programs. By demonstrating measurable outcomes through data, organizations can appeal to donors who want to support programs with proven impact. Major philanthropic foundations, family trusts, and high-net-worth individuals increasingly demand data transparency in their giving.

4. Social Impact Bonds (SIBs)

Social Impact Bonds are an innovative model that allows private investors to fund social programs with the potential for government payback based on measurable outcomes. If a homelessness program funded by an SIB successfully reduces homelessness rates, the government repays the investors. This model aligns public and private interests, encouraging investment in sustainable solutions.

5. Corporate Social Responsibility (CSR)

Corporations are increasingly committing funds to social causes as part of their CSR initiatives. By partnering with companies, particularly those in the tech and data sectors, homelessness programs can receive both funding and technological support, helping them operate more efficiently and reach more people.

Innovative Funding Through Data-Driven Approaches

Data-driven results make programs more attractive to funders who value transparency and measurable impact. Funding agencies, foundations, and private donors are often more inclined to invest when they can see clear evidence of a program's effectiveness. For example, by tracking the progress of individuals in job placement programs and analyzing the impact on their housing stability, an organization can demonstrate its role in reducing homelessness.

Impact investors are another potential funding source that appreciates data transparency. Unlike traditional donors, impact investors seek both social and financial returns. When homelessness programs can provide data showing a clear return on social impact, impact investors are more likely to provide funding, knowing that their investment is creating measurable positive change.

Building Partnerships for Ongoing Support

Sustainability is not achieved in isolation. Partnerships between government agencies, nonprofits, private companies, and community organizations amplify the reach and impact of homelessness programs. These partnerships allow for resource sharing, cost reduction, and increased access to services.

Cross-sector partnerships are particularly effective for homelessness prevention, as they bring together a variety of expertise and resources. For example, a partnership between a tech company and a homelessness prevention nonprofit could result in the development of predictive analytics tools that help identify those at risk of becoming homeless. Likewise, collaboration with healthcare providers can improve access to mental health and addiction services, addressing root causes of homelessness and reducing the likelihood of recurrence.

Integrating Community Input for Sustainable Impact

Community involvement is essential for sustainable change. By engaging the people most affected by homelessness, programs can gain insights into local needs and tailor solutions accordingly. This involvement also fosters community ownership, increasing the likelihood of long-term success.

Incorporating community feedback ensures that programs are meeting actual needs rather than perceived ones. Surveys, public forums, and advisory councils of formerly homeless individuals offer valuable perspectives, helping funders understand the realities on the ground and design initiatives that address core issues. Data can be an effective tool here as well, helping communicate community needs to funders in a compelling way.

Monitoring, Evaluation, and Accountability

Regular monitoring and evaluation are critical for sustainable impact. By tracking key metrics—such as reductions in homelessness rates, job placements, and improvements in quality of life—programs can ensure they remain effective and make adjustments as needed.

Transparent reporting mechanisms, like annual reports and public dashboards, are essential for accountability. By sharing data-driven results with stakeholders, organizations demonstrate their commitment to achieving measurable outcomes. This transparency not only helps maintain funder confidence but also builds trust within the community, fostering greater support for the program.

Securing Long-Term Funding Commitments

Long-term funding commitments provide stability, allowing programs to plan and operate more effectively. By consistently demonstrating impact, organizations can secure multi-year commitments from funders, reducing the uncertainty associated with year-to-year funding. Reliable funding allows programs to retain experienced staff, develop deeper community relationships, and invest in strategic long-term planning.

Programs that secure long-term commitments are often better able to address systemic issues that contribute to homelessness, as they have the resources to look beyond immediate needs and implement changes at a structural level. By aligning data with funders' goals, organizations can make a compelling case for ongoing support.

Scaling and Replicating Success in Other Cities

A data-driven model that achieves sustainability in one city can serve as a blueprint for addressing homelessness in other cities and countries. While the specifics of homelessness may vary by location, many of the underlying issues—such as lack of affordable housing, mental health challenges, and economic instability—are universal. Data-driven solutions are adaptable, responding to local needs while remaining rooted in proven strategies.

Los Angeles, with its diverse population and unique housing market challenges, serves as an ideal testing ground for this model. If a sustainable approach can succeed here, it could be scaled to cities across the U.S. and potentially around the world. By using data to customize solutions for each new location, the impact of this model could be far-reaching, extending hope and support to millions.

Sustainability in homelessness prevention demands commitment—from funders, policymakers, nonprofits, corporations, and individuals. Each has a role to play in ensuring that homelessness programs are not only adequately funded but also strategically designed to achieve lasting change. By combining data with funding, partnerships, and community input, we can create a sustainable model for ending homelessness.

This is more than a book about data; it's a call to action. Let's harness the power of data and turn it into stories, strategies, and real solutions that last.

CHAPTER 22: COMMUNITY ENGAGEMENT: EMPOWERING LOCAL COMMUNITIES

Addressing homelessness in Los Angeles is not just a matter of policy and resources but also requires the active involvement and empowerment of local communities.

C ommunity engagement is essential for creating tailored, effective solutions that address the specific needs of each neighborhood. This chapter explores strategies for involving community members, fostering local leadership, and leveraging community resources to make a lasting impact.

The Importance of Community Engagement in Tackling Homelessness

Community engagement fosters a sense of shared responsibility and partnership in addressing homelessness. When communities actively participate in creating solutions, the resulting initiatives are more likely to reflect local values, gain public support, and lead to sustainable outcomes.

Benefits of Community Engagement:

Building Trust: Establishing trust between service providers, policymakers, and community members.

Local Insights: Leveraging the unique perspectives and experiences of residents to create relevant solutions.

Increased Accountability: Encouraging transparency and community oversight, which strengthens accountability.

Understanding Community Needs Through Participatory Research

Engaging communities in the research process helps ensure that data accurately reflects their needs and priorities. Participatory research methods, such as community surveys, focus groups, and town hall meetings, allow community members to share their insights, shaping the direction of homelessness initiatives.

Participatory Research Techniques:

Community Surveys: Collect data on local experiences, concerns, and ideas.

Focus Groups: Facilitate discussions with diverse community members to understand a broad range of perspectives.

Community-Based Advisory Boards: Establish boards with local representatives who can provide ongoing input.

Empowering Local Leaders as Change Agents

Local leaders, including community organizers, faith-based leaders, and neighborhood advocates, play a critical role in mobilizing support for homelessness initiatives. By empowering these leaders, organizations can foster a culture of support and shared responsibility at the grassroots level.

Strategies for Empowering Leaders:

Training and Resources: Provide leaders with the tools and resources they need to engage effectively, including workshops on advocacy and homelessness education.

Collaboration Opportunities: Offer platforms for local leaders to collaborate with service providers, policymakers, and other stakeholders.

Recognition and Support: Recognize the efforts of local leaders and provide continuous support to sustain their involvement.

Creating Community-Driven Solutions

Solutions that are created and driven by community members tend to be more sustainable and impactful. By empowering residents to co-design and implement homelessness initiatives, organizations can ensure that solutions reflect local priorities and are better accepted within the community.

Community-Driven Approaches:

Neighborhood Action Plans: Develop tailored action plans that address specific needs in different neighborhoods.

Peer Support Networks: Encourage the creation of peer-led support groups to provide mutual assistance.

Community Resource Hubs: Set up community centers or information hubs where residents can access resources and services.

Building Awareness and Reducing Stigma

Homelessness is often surrounded by stigma and misunderstanding. Educating communities about the causes of homelessness and the potential for recovery can foster compassion and reduce barriers to effective engagement.

Awareness Campaigns:

Public Education: Use local media, social media, and community events to share information and reduce stigma.

Humanizing Homelessness: Share personal stories and testimonials from those who have experienced homelessness to promote empathy.

School Outreach Programs: Engage youth through educational programs that teach empathy and social responsibility.

Engaging the Broader Community in Volunteering and Support

Volunteering offers community members a hands-on way to contribute to homelessness solutions. Engaging volunteers not only expands the reach of homelessness services but also helps volunteers understand the complexity of homelessness, fostering a culture of compassion.

Volunteer Programs:

Outreach Teams: Engage local volunteers in outreach efforts, such as distributing meals or providing companionship.

Skill-Based Volunteering: Encourage community members to volunteer their professional skills, such as counseling, healthcare, or legal assistance.

Fundraising Events: Organize community-driven events to raise funds and awareness for homelessness programs.

The Role of Faith-Based Organizations and Nonprofits

Faith-based organizations and nonprofits have a longstanding tradition of supporting homeless individuals through charitable work. Partnering with these groups can enhance community-based solutions, creating a wider safety net for individuals at risk of or experiencing homelessness.

Faith-Based and Nonprofit Partnerships:

Service Collaborations: Partner with these organizations to offer services like food distribution, emergency shelters, and counseling.

Community Outreach: Leverage these groups' established relationships within the community to reach vulnerable populations.

Advocacy Efforts: Work together to advocate for policy changes that benefit the homeless population.

Engaging Local Businesses and Private Sector Support

Businesses, both small and large, can contribute meaningfully to homelessness solutions through funding, resources, and job opportunities. Engaging local businesses fosters economic development and encourages the private sector to invest in community well-being.

Private Sector Initiatives:

Corporate Social Responsibility (CSR) Programs: Encourage businesses to allocate CSR funds toward homelessness programs.

Employment Opportunities: Collaborate with businesses to create job opportunities for individuals transitioning out of homelessness.

Donations and Sponsorships: Engage businesses in donating goods or sponsoring programs.

Measuring the Impact of Community Engagement

Evaluating the impact of community engagement efforts ensures that strategies are effective and helps identify areas for improvement. By measuring outcomes, organizations can adapt and refine their approaches to maximize the positive impact on homelessness.

Impact Measurement Approaches:

Qualitative Feedback: Gather feedback from community members to assess satisfaction and engagement levels.

Quantitative Metrics: Track metrics, such as participation rates, resource distribution, and volunteer hours.

Long-Term Outcomes: Measure the long-term impact of

community engagement on homelessness rates and service accessibility.

Creating a Sustainable Culture of Community Involvement

To sustain the fight against homelessness, it is essential to create a culture of ongoing involvement where community members feel empowered to contribute. Building this culture requires regular engagement, transparency, and recognition of community contributions.

Strategies for Sustaining Involvement:

Regular Communication: Maintain open communication channels to keep the community informed and involved.

Celebrating Successes: Celebrate milestones and achievements with the community to reinforce commitment.

Continuous Engagement Opportunities: Offer ongoing opportunities for community members to participate in and support homelessness initiatives.

Empowering Communities to Drive Change

Empowering communities is vital for developing sustainable, localized solutions to homelessness in Los Angeles. By involving residents, leaders, businesses, and organizations, homelessness initiatives can benefit from a diverse array of insights, resources, and support. When communities feel a sense of ownership and responsibility, they are more likely to invest in lasting change.

Through community engagement, the fight against homelessness becomes a collective effort, bringing together individuals from all walks of life. By fostering empathy, reducing stigma, and empowering local leaders, communities can create a supportive environment where everyone is committed to making homelessness a thing of the past.

CHAPTER 23: CONCLUSION: PREDICTING HOPE FOR A BRIGHTER FUTURE

Insights,strategies and possibilities explored.

In this final chapter, we bring together the insights, strategies, and possibilities explored throughout the book, emphasizing the role of data in not just predicting outcomes but in fostering hope and actionable change for those experiencing homelessness in Los Angeles. As homelessness remains one of the most complex social issues facing urban centers, the progress made through data-driven approaches reveals a path forward—one where informed interventions, community collaboration, and technological innovation work hand-in-hand to create a future with fewer people experiencing homelessness and more finding pathways to stability and dignity.

This chapter concludes the book and serves as a powerful call to action, reminding readers of the impact that data-driven approaches can have in addressing homelessness and inspiring them to contribute to this mission. By acknowledging the role of everyone—from policymakers to community members—in this journey, the conclusion reinforces the potential for a brighter future, grounded in both empathy and evidence.

Reflections on the Power of Data in Social Change

The journey through each chapter of this book illustrates how data, when used ethically and strategically, can transcend mere numbers to become a powerful tool for empathy, understanding, and transformation. Data enables policymakers, advocates, and community organizations to see beyond individual stories and spot trends, identify effective interventions, and allocate resources where they are most needed.

Key Takeaways:

Data as Insight: Transforming raw data into meaningful insights that can inform policies and drive results.

Data as Empathy: By revealing the systemic challenges faced by individuals, data allows us to approach solutions with compassion and urgency.

Data as Action: When turned into actionable strategies, data enables us to move from insight to tangible improvements in people's lives.

The Role of Innovation and Emerging Technologies

Emerging technologies, from predictive analytics to geospatial mapping, continue to open new possibilities for understanding and addressing homelessness. With advances in artificial intelligence, machine learning, and mobile data collection, cities like Los Angeles are poised to harness even more precise and effective tools for reducing homelessness.

Future Innovations:

AI and Machine Learning: Enhanced predictive models that can help prevent homelessness before it starts.

Mobile and Cloud-Based Platforms: Increasing accessibility to data in real-time, supporting rapid responses and emergency interventions.

Blockchain and Data Security: Securing personal information while enabling inter-agency collaboration and accountability.

A Vision for Lasting Change: Scaling and Replicating Success

The strategies and models developed in Los Angeles can serve as a blueprint for cities worldwide facing similar challenges. By sharing insights, best practices, and case studies, LA can become a model city for tackling homelessness through data-driven approaches.

Replicating Success:

Adapting LA's Model: Helping other cities implement data-based solutions customized to their unique needs.

Building a National Framework: Creating standards and protocols for data-driven homelessness initiatives across the U.S.

Global Partnerships: Engaging with cities around the world to exchange knowledge and support international efforts to address homelessness.

Strengthening Community Trust and Empowerment

A brighter future requires the sustained engagement and empowerment of local communities. By involving residents in data collection, analysis, and solutions, cities can create a more inclusive approach that recognizes the role of each individual in creating change.

Building a Culture of Inclusion:

Transparent Communication: Sharing data findings and decisions openly with the public to build trust.

Empowering Communities: Training community leaders and members in data literacy and advocacy.

Ongoing Engagement: Creating opportunities for communities to be active participants, not just recipients of solutions.

The Path Forward: Policy, Funding, and Long-Term Impact

Sustaining data-driven solutions requires supportive policies, consistent funding, and a commitment to measuring long-term outcomes. Policymakers, advocates, funders, and community members must work together to maintain the momentum and address emerging challenges.

Policy Recommendations:

Investment in Data Infrastructure: Ensuring that resources are allocated to maintain and improve data systems.

Flexible and Responsive Policies: Adapting policies as data reveals new insights or emerging needs.

Funding for Long-Term Solutions: Committing to sustained funding that goes beyond temporary relief to create lasting change.

FINAL THOUGHTS: PREDICTING HOPE

Predicting hope is not about guaranteeing an end to homelessness overnight but about envisioning a future where data-driven actions reduce homelessness, support those in need, and foster a society where every person can thrive. By continuing to use data as both a guiding tool and a source of hope, we create a vision for a Los Angeles—and a world—where homelessness becomes a rare exception, not the rule.

Through collaboration, empathy, and innovation, the collective efforts of government agencies, nonprofits, private sectors, and communities have the potential to make a lasting impact. Together, we can move beyond merely addressing homelessness to preventing it and, ultimately, ending it.

Epilogue: A Call to Action

The journey to end homelessness in Los Angeles begins with each of us. Whetheryou are a policymaker, a data scientist, an advocate, or simply a concerned resident, your role is crucial. By supporting data-driven initiatives, advocating for fair policies, or volunteering time and resources, every effort counts.

As a Ghanaian author now residing in Los Angeles, the writer brings a unique cultural lens that values community cohesion, extended family support systems, and collective responsibility—elements deeply embedded in Ghanaian society. This perspective not only enhances the author's understanding of homelessness but also allows for a fresh viewpoint in addressing the complex issues around it in the U.S. His background offers potential value for American society as a whole, especially in a diverse city like Los Angeles, which faces a unique and challenging homelessness crisis. By integrating Ghanaian principles of community-centered support and culturally relevant approaches, the author can offer new solutions that resonate with both local and diverse populations.

In Ghana, homelessness is approached through a lens that emphasizes collective responsibility, where extended family and community networks often step in to prevent individuals from falling into destitution. This foundational belief is one the author believes could benefit American society in addressing homelessness. Unlike the prevalent individualistic approach in many Western societies, which tends to prioritize personal responsibility, Ghanaian values encourage shared responsibility and a communal safety net. By introducing these concepts, the author seeks to inspire American policymakers, social workers, and community leaders to consider more community-driven approaches, such as mentorship programs or "sponsor families," to support individuals at risk of homelessness.

Furthermore, Ghana's community savings groups and cooperative housing initiatives, which allow low-income families to collectively pool resources for housing or mutual support, offer a model that could resonate in the context of Los Angeles. The author believes that by adapting these financial models, American society could create more inclusive housing options for the homeless and low-income populations, perhaps even fostering a sense of community ownership and pride. This could prove particularly valuable in LA, where the housing crisis has deepened due to high costs and limited affordable housing options. Such initiatives could reduce housing costs and offer a sustainable support network, potentially alleviating some of the city's housing challenges.

The author's experience with culturally sensitive approaches to mental health in Ghana also brings a fresh perspective to LA's homelessness crisis. By valuing local traditions alongside mental health interventions, Ghanaian mental health strategies blend Western and indigenous practices. This approach, the author believes, could be highly beneficial in LA, especially in communities with diverse backgrounds where culturally relevant mental health support can foster stronger trust and engagement. Such tailored approaches can be instrumental in providing effective mental health services to homeless populations, particularly those who feel marginalized or disconnected from mainstream support systems.

Additionally, the author advocates for early intervention and preventive measures—a priority in Ghanaian social policy. This philosophy aligns with the concept of engaging at-risk individuals before they fall into homelessness, rather than waiting until they require intensive support. By focusing on community-level outreach and prevention, the author envisions creating sustainable pathways for reducing homelessness in LA. Initiatives aimed at families and youth could provide stability, offering a layer of protection and resilience to those on the edge of homelessness.

Finally, the author's Ghanaian heritage fosters a holistic view of data and policy, wherein policies are not only informed by quantitative data but also grounded in qualitative, community-driven insights. The author believes this dual approach can bring immense value to Los Angeles. While LA's current systems rely heavily on quantitative data, incorporating community-centered insights could offer a richer, more human understanding of homelessness. Data could then be used not just to identify risks but also to understand the community values, beliefs, and practices that shape people's lives and struggles.

The author's Ghanaian background therefore brings unique insights to American society, offering strategies that prioritize community resilience, cultural sensitivity, and preventive care. In Los Angeles, these approaches could prove especially beneficial, aligning with the city's diversity and complexity, and fostering a more inclusive, compassionate, and effective response to homelessness. Through these contributions, the author hopes to bridge Ghanaian values and American innovation, ultimately enhancing LA's capacity to address homelessness and supporting a broader dialogue on community and social responsibility across the U.S.

Ending homelessness is an ambitious goal that requires more than just government action; it demands a united community. It calls for a shared commitment to both compassion and evidence-based solutions. As individuals, we can champion the cause by pushing for accountability, investing in long-term initiatives, and ensuring the voices of those affected by homelessness are heard and respected.

Together, by blending innovation with compassion, we can predict a future where hope outshines despair, where every person has access to a safe home, and where homelessness is no longer an unsolvable crisis but a chapter that has been closed in our shared history.

Let this book be more than just a guide; let it be an invitation to join the journey toward lasting, positive change.

ISAAC GYAN

GLOSSARY

1. Affordable Housing – Housing that is reasonably priced, allowing low-income households to meet basic living costs.

2. At-Risk Population – Groups more likely to face adverse outcomes, such as homelessness, due to factors like poverty or mental illness.

3. Behavioral Health – An umbrella term for mental health, substance abuse, and their effects on wellbeing.

4. Benchmarking – Setting performance standards based on best practices to evaluate effectiveness.

5. Big Data – Large and complex datasets that require advanced tools for processing and analysis.

6. Census Data – Demographic data collected by the government to provide population insights.

7. Chronic Homelessness – Long-term homelessness, often involving repeated episodes and linked to disabling conditions.

8. Community-Based Organization (CBO) – Nonprofits or other entities rooted in local communities working to address issues like homelessness.

9. Continuum of Care (CoC) – A coordinated approach to provide homeless services from outreach to long-term housing.

10. Data-Driven Decision Making (DDDM) – Using data analysis to guide and inform policy and practice decisions.

11. Data Ethics – Principles governing the responsible and fair use of data, including privacy and consent.

12. Data Literacy – The ability to read, interpret, and use data effectively.

13. Demographic Analysis – Examination of population statistics, such as age and ethnicity, to understand social patterns.

14. Discharge Planning – Preparing individuals for stable housing and support upon release from institutions like hospitals or prisons.

15. Emergency Shelter – Temporary accommodations for people experiencing homelessness.

16. Evidence-Based Practices – Practices grounded in data and research showing proven outcomes.

17. Geospatial Analysis – Mapping data related to geographic locations, useful for identifying service gaps.

18. Homeless Management Information System (HMIS) – A data system used by homeless service providers to track client data and service needs.

19. Homelessness Prevention – Interventions designed to prevent individuals from becoming homeless.

20. Housing First – A model that prioritizes immediate housing for homeless individuals without preconditions.

21. Longitudinal Study – Research that follows subjects over a long period to observe changes and outcomes.

22. Machine Learning (ML) – A form of artificial intelligence enabling computers to learn patterns from data.

23. Marginalized Communities – Groups facing disadvantages due to systemic barriers, such as racial or economic inequities.

24. Mental Health Services – Programs aimed at providing therapy, medication, or other support for mental health needs.

25. Metrics – Quantitative measures used to assess program

success and outcomes.

26. Outreach Programs – Efforts by service providers to connect with individuals on the streets or in shelters.

27. PIT Count (Point-in-Time Count) – An annual count of the homeless population on a single night, usually conducted in January.

28. Predictive Analytics – Using historical data to forecast future events or outcomes, like homelessness risk.

29. Public Policy – Laws and regulations enacted by governments to address societal issues.

30. Recidivism – The tendency of formerly homeless individuals to return to homelessness.

31. Rent Burden – When housing costs consume a high percentage of a household's income, increasing vulnerability to homelessness.

32. Resource Allocation – Distribution of resources, such as funding or services, to areas of greatest need.

33. Risk Assessment – Evaluation of factors that increase an individual's likelihood of experiencing homelessness.

34. Scatter-Site Housing – Dispersed housing units integrated into various neighborhoods, often part of Housing First programs.

35. Shelter Utilization Rate – The percentage of available shelter beds that are occupied.

36. Social Determinants of Health – Economic and social conditions influencing health outcomes, including housing stability.

37. Sociodemographic Factors – Characteristics like age, gender, income, and education that impact an individual's life situation.

38. Stakeholder – Individuals or groups with an interest in

addressing homelessness, including policymakers, nonprofits, and residents.

39. Substance Abuse Disorder – Addiction-related conditions impacting physical and mental health.

40. Supportive Housing – Housing paired with supportive services to help individuals maintain stability.

41. Sustainability – The ability to maintain or replicate programs long-term without exhausting resources.

42. Systemic Inequity – Institutional practices that result in unequal access to resources and opportunities for certain groups.

43. Transitional Housing – Temporary housing that helps individuals move from homelessness to permanent housing.

44. Trauma-Informed Care – An approach that recognizes the impact of trauma on individuals and adapts services to accommodate.

45. Underhoused – Individuals or families living in inadequate, overcrowded, or insecure housing.

46. Unsheltered Homelessness – People living on the streets or in places unfit for human habitation.

47. Urban Poverty – Poverty that occurs in urban settings, often associated with higher costs of living and fewer affordable housing options.

48. Veteran Homelessness – Homelessness affecting former military personnel, often with unique causes and needs.

49. Wraparound Services – Comprehensive services addressing multiple needs, including housing, mental health, and job training.

50. Zero Homelessness – A goal or state in which homelessness has been eliminated, with systems in place to prevent its recurrence.

REFERENCE

Books

1. Burt, M. R. (2001). Helping America's homeless: Emergency shelter or affordable housing? The Urban Institute Press.

2. Culhane, D. P., & Metraux, S. (2008). A public health approach to ending homelessness. Journal of Public Health, 98(5), 781-787.

3. Elliott, M. (2014). The sociology of homelessness: Ethnographic studies of the homeless. Routledge.

4. Fitzpatrick, S., & Christian, J. (2006). The hidden truth about homelessness: Experiences of single homelessness in England. Crisis UK.

5. Gaetz, S., & Dej, E. (2017). A new direction: A framework for homelessness prevention. Canadian Observatory on Homelessness Press.

6. Goodman, L. A., & Thomson, D. M. (2021). The ethical use of predictive analytics in social work. NASW Press.

7. Grigsby, C., Baumann, D., Gregorich, S. E., & Roberts-Gray, C. (1990). Disaffiliation to entrenchment: A model for understanding homelessness. Journal of Social Issues, 46(4), 141-156.

8. Henry, M., Watt, R., Mahathey, A., Ouellette, J., & Sitler, A. (2019). The 2019 Annual Homeless Assessment Report (AHAR) to Congress. HUD.

9. Hopper, K., & Baumohl, J. (2005). Redefining the cursed word: A historical interpretation of American homelessness. Routledge.

10. Johnson, G., & Chamberlain, C. (2011). Are the homeless mentally ill? A case study of homelessness in Australia. Routledge.

11. Kuhn, R., & Culhane, D. P. (1998). Applying cluster analysis to test a typology of homelessness by pattern of shelter utilization: Results from the analysis of administrative data. American Journal of Community Psychology, 26(2), 207-232.

12. Link, B. G., & Phelan, J. (2001). Conceptualizing stigma. Annual Review of Sociology, 27, 363-385.

13. Mayock, P., & Bretherton, J. (2016). Women's homelessness in Europe. Palgrave Macmillan.

14. Metraux, S., & Culhane, D. P. (1999). Family dynamics, housing stability, and homelessness. The Urban Institute.

15. O'Flaherty, B. (2010). How to house the homeless. Russell Sage Foundation.

16. Rossi, P. H. (1989). Down and out in America: The origins of homelessness. University of Chicago Press.

17. Shinn, M., & Khadduri, J. (2020). In the midst of plenty: Homelessness and what to do about it. Wiley.

18. Speer, J. (2018). Cities, police, and homelessness: Using theory to understand homelessness policy and street-level repression. Routledge.

19. Wright, T. (2016). Out of place: Homeless mobilizations, subcities, and contested landscapes. SUNY Press.

20. Wolch, J., & Dear, M. (2014). Landscapes of despair: From deinstitutionalization to homelessness. Princeton University Press.

Scholarly Articles

1. Baggett, T. P., O'Connell, J. J., Singer, D. E., & Rigotti, N.

A. (2010). The unmet health care needs of homeless adults: A national study. American Journal of Public Health, 100(7), 1326-1333. https://doi.org/10.2105/AJPH.2009.180109

2. Barrow, S., Herman, D., Córdova, P., & Struening, E. (1999). Mortality among homeless shelter residents in New York City. American Journal of Public Health, 89(4), 529-534. https://doi.org/10.2105/AJPH.89.4.529

3. Bassuk, E. L., & Geller, S. (2006). The role of housing and services in ending family homelessness. Housing Policy Debate, 17(4), 781-806. https://doi.org/10.1080/10511482.2006.9521580

4. Byrne, T., Treglia, D., Culhane, D. P., Kuhn, J., & Kane, V. (2016). Predictors of homelessness among families and single adults after exit from homelessness prevention and rapid re-housing programs: Evidence from the Department of Veterans Affairs Supportive Services for Veteran Families Program. Housing Policy Debate, 26(1), 252-275. https://doi.org/10.1080/10511482.2015.1041257

5. Caton, C. L., Wilkins, C., & Anderson, J. (2007). People who experience long-term homelessness: Characteristics and interventions. Journal of Urban Health, 84(4), 524-534. https://doi.org/10.1007/s11524-007-9178-9

6. Culhane, D. P. (2008). The cost of homelessness: A perspective from the United States. European Journal of Homelessness, 2(1), 97-114.

7. Desmond, M., & Kimbro, R. T. (2015). Eviction's fallout: Housing, hardship, and health. Social Forces, 94(1), 295-324. https://doi.org/10.1093/sf/sov044

8. Elbogen, E. B., Sullivan, C., Wolfe, J., Wagner, H. R., & Beckham, J. C. (2013). Homelessness and money mismanagement in Iraq and Afghanistan veterans. American Journal of Public Health, 103(2), 248-254. https://doi.org/10.2105/AJPH.2013.301322

9. Fazel, S., Geddes, J. R., & Kushel, M. (2014). The health of homeless people in high-income countries: Descriptive epidemiology, health consequences, and clinical and policy recommendations. The Lancet, 384(9953), 1529-1540. https://doi.org/10.1016/S0140-6736(14)61132-6

10. Greenberg, G. A., & Rosenheck, R. A. (2008). Jail incarceration, homelessness, and mental health: A national study. Psychiatric Services, 59(2), 170-177. https://doi.org/10.1176/ps.2008.59.2.170

11. Henwood, B. F., & Katz, M. L. (2015). Housing stability, physical health status, and health care utilization among homeless adults in Los Angeles. Journal of Urban Health, 92(4), 785-795. https://doi.org/10.1007/s11524-015-9970-6

12. Hopper, E. K., Bassuk, E. L., & Olivet, J. (2010). Shelter from the storm: Trauma-informed care in homelessness services settings. The Open Health Services and Policy Journal, 3, 80-100. https://doi.org/10.2174/1874924001003020080

13. Johnsen, S., & Fitzpatrick, S. (2010). Revanchist sanitization or coercive care? The use of enforcement to combat street homelessness in the UK and USA. Urban Studies, 47(8), 1703-1723. https://doi.org/10.1177/0042098009356128

14. Jones, M. M. (2015). Does race matter in addressing homelessness? A review of the literature. World Medical & Health Policy, 7(1), 121-136. https://doi.org/10.1002/wmh3.132

15. Kuhn, R., & Culhane, D. P. (1998). Applying cluster analysis to test a typology of homelessness by pattern of shelter utilization: Results from the analysis of administrative data. American Journal of Community Psychology, 26(2), 207-232. https://doi.org/10.1023/A:1022176402357

16. Lee, B. A., Tyler, K. A., & Wright, J. D. (2010). The new homelessness revisited. Annual Review of Sociology, 36, 501-521. https://doi.org/10.1146/annurev-

soc-070308-115940

17. Metraux, S., & Culhane, D. P. (2006). Homeless shelter use and reincarceration following prison release: Assessing the risk. Criminology & Public Policy, 5(2), 185–208. https://doi.org/10.1111/j.1745-9133.2006.00376.x

18. Padgett, D. K., Stanhope, V., Henwood, B. F., & Stefancic, A. (2011). Substance use outcomes among homeless clients with serious mental illness: Comparing Housing First with Treatment First programs. Community Mental Health Journal, 47(2), 227-232. https://doi.org/10.1007/s10597-009-9283-7

19. Pearson, C. L., Montgomery, A. E., & Locke, G. (2009). Housing stability among homeless individuals with serious mental illness participating in Housing First programs. Journal of Community Psychology, 37(3), 404-417. https://doi.org/10.1002/jcop.20301

20. Petrovich, J., & Robbins, D. L. (2017). Predictors of homelessness among veterans: A systematic review. Social Work in Public Health, 32(6), 412-429. https://doi.org/10.1080/19371918.2017.1316676

21. Pleace, N. (2016). Exclusion by definition: The underrepresentation of women in European homelessness statistics. European Journal of Homelessness, 10(1), 163-183.

22. Raoult, D., & Arzounian, D. (2012). Health care needs of the homeless population. International Journal of Environmental Research and Public Health, 9(4), 941-956. https://doi.org/10.3390/ijerph9040941

23. Rog, D. J., & Buckner, J. C. (2007). Homeless families and children. In D. Dennis, G. Locke, & J. Khadduri (Eds.), Toward understanding homelessness: The 2007 national symposium on homelessness research (pp. 1-32). U.S. Department of Housing and Urban Development.

24. Rosenheck, R., & Fontana, A. (1994). A model of

homelessness among male veterans of the Vietnam War generation. American Journal of Psychiatry, 151(3), 421-427. https://doi.org/10.1176/ajp.151.3.421

25. Shlay, A. B., & Rossi, P. H. (1992). Social science research and contemporary studies of homelessness. Annual Review of Sociology, 18, 129-160. https://doi.org/10.1146/annurev.so.18.080192.001021

26. Stergiopoulos, V., Gozdzik, A., Nisenbaum, R., & Durbin, J. (2015). Effectiveness of Housing First programs in retaining homeless people with mental illness. Health Affairs, 34(7), 1166-1173. https://doi.org/10.1377/hlthaff.2014.1528

27. Tsai, J., & Rosenheck, R. A. (2015). Risk factors for homelessness among US veterans. Epidemiologic Reviews, 37(1), 177-195. https://doi.org/10.1093/epirev/mxu004

28. Tsemberis, S. (2010). Housing First: Ending homelessness, promoting recovery, and reducing costs. In I. Gould Ellen & B. O'Flaherty (Eds.), How to house the homeless (pp. 37-56). Russell Sage Foundation.

29. Turner, N., & Moran, L. (2020). Using predictive analytics to address homelessness. Journal of Social Service Research, 46(3), 325-338. https://doi.org/10.1080/01488376.2020.1717148

30. Williams, A. M., & Hall, C. M. (2014). Homelessness, health, and policy: A European perspective. European Journal of Public Health, 24(2), 169-176. https://doi.org/10.1093/eurpub/ckt101

ABOUT THE AUTHOR

Isaac Gyan is a dynamic and results-driven Business Analyst, with over 11 years of experience across Banking, Finance, Marketing, and Research Data Analysis. With a passion for transforming data into actionable insights, Isaac has dedicated his career to optimizing business processes and driving impactful organizational improvements. He is currently a Doctoral candidate in Business Administration, specializing in Business Intelligence and Data Analytics, at Westcliff University in Irvine, California. Isaac's extensive academic background includes a Master's degree in International Business from the University of Verona and a Master's in Economics and Entrepreneurship from the University of Cassino and Southern Lazio.

Over the years, Isaac has worked with various organizations, applying his expertise in data analytics, statistical analysis, and business intelligence to create innovative solutions that have improved market strategies, optimized financial operations, and enhanced decision-making processes. As a market analyst at Latex Foam Ltd and a banking operations analyst at Akuapem Rural Bank in Ghana, he gained valuable experience in using data to understand consumer behavior, identify market trends, and develop strategies for business growth.

Isaac's work is grounded in the belief that data holds the power to unlock solutions to some of society's most pressing challenges. In Predicting Hope: Utilizing Data to End Hopelessness in LA, Isaac combines his expertise in business analytics with his passion for social change. Drawing on his extensive experience in data collection, processing, and analysis, Isaac explores how data-driven insights can be used to tackle the issue of hopelessness in Los Angeles. His goal is to provide practical, research-backed solutions for empowering communities, improving lives, and transforming the future.

Isaac currently resides in Los Angeles, where he continues to study, research, and work on innovative data solutions that aim to improve lives and create lasting change. His passion for utilizing technology and data in meaningful ways is central to his mission of making the world a better place, one insight at a time.

www.ingramcontent.com/pod-product-compliance
Lightning Source LLC
LaVergne TN
LVHW051225050326
832903LV00028B/2250